550 AP®

BIOLOGY
Practice Questions

The Staff of the Princeton Review

PrincetonReview.com

PENGUIN RANDOM HOUSE

The Princeton Review, Inc.
24 Prime Parkway, Suite 201
Natick, MA 01760
E-mail: editorialsupport@review.com

Published in the United States by Random House, Inc., New York, and in
Canada by Random House of Canada Limited, Toronto.
A Penguin Random House Company.

The Princeton Review is not affiliated with Princeton University.

AP and Advanced Placement Program are registered trademarks of the
College Board which does not sponsor or endorse this product.

ISBN: 978-0-8041-2488-1
ISSN: 2330-7013

Editor: Calvin S. Cato
Production Editor: Liz Rutzel
Production Artist: Deborah A. Silvestrini

Printed in the United States of America on partially
recycled paper.

10 9 8 7 6 5 4 3 2 1

Editorial

Rob Franek, Senior VP, Publisher
Casey Cornelius, VP, Content Development
Mary Beth Garrick, Director of Production
Selena Coppock, Managing Editor
Calvin Cato, Editor
Colleen Day, Editor
Aaron Riccio, Editor
Meave Shelton, Editor
Alyssa Wolff, Editorial Assistant

Random House Publishing Team

Tom Russell, Publisher
Alison Stoltzfus, Publishing Manager
Dawn Ryan, Associate Managing Editor
Ellen Reed, Production Manager
Erika Pepe, Associate Production Manager
Kristin Lindner, Production Supervisor
Andrea Lau, Designer

Acknowledgments

The Princeton Review would like to give a very special thanks to Mary DeAgostino, Christopher Stobart, Keith Syska, and Sarah Ann Woodruff for their hard work on the creation of this title. In addition, The Princeton Review would like to thank Liz Rutzel for her diligence in copy-editing this title and Deborah Silvestrini for her hard work on the production of this book.

Contents

Part I: Using This Book to Improve Your AP Score 1

    Preview: Your Knowledge, Your Expectations 2

    Your Guide to Using This Book.. 2

    How to Begin.. 3

    AP Biology Diagnostic Test 7

    AP Biology Diagnostic Test Answers and Explanations.................. 31

Part II: About the AP Biology Exam.. 43

    The Structure of the AP Biology Exam...................................... 44

    Overview of Content Topics 44

    How AP Exams Are Used 46

    Other Resources 47

    Designing Your Study Plan 47

Part III: Test-Taking Strategies for the AP Biology Exam 49

  1 How to Approach Multiple-Choice Questions 51

  2 How to Approach Free-Response Questions............................. 61

Part IV: Drills... 71

  3 Big Idea 1 Drill 1 73

  4 Big Idea 1 Drill 1 Answers and Explanations............................. 81

  5 Big Idea 2 Drill 1 87

  6 Big Idea 2 Drill 1 Answers and Explanations............................. 95

  7 Big Idea 3 Drill 1 101

  8 Big Idea 3 Drill 1 Answers and Explanations............................. 109

  9 Big Idea 4 Drill 1 115

 10 Big Idea 4 Drill 1 Answers and Explanations............................. 125

 11 Big Idea 1 Drill 2 131

 12 Big Idea 1 Drill 2 Answers and Explanations............................. 139

 13 Big Idea 2 Drill 2 145

 14 Big Idea 2 Drill 2 Answers and Explanations............................. 153

 15 Big Idea 3 Drill 2 159

 16 Big Idea 3 Drill 2 Answers and Explanations............................. 167

 17 Big Idea 4 Drill 2 173

 18 Big Idea 4 Drill 2 Answers and Explanations............................. 181

 19 Big Idea 1 Drill 3 187

 20 Big Idea 1 Drill 3 Answers and Explanations............................. 197

 21 Big Idea 2 Drill 3 205

 22 Big Idea 2 Drill 3 Answers and Explanations............................. 215

 23 Big Idea 3 Drill 3 223

24 Big Idea 3 Drill 3 Answers and Explanations 231

25 Big Idea 4 Drill 3 239

26 Big Idea 4 Drill 3 Answers and Explanations 249

27 Big Idea 1 Drill 4 257

28 Big Idea 1 Drill 4 Answers and Explanations 263

29 Big Idea 2 Drill 4 267

30 Big Idea 2 Drill 4 Answers and Explanations 277

31 Big Idea 3 Drill 4 283

32 Big Idea 3 Drill 4 Answers and Explanations 291

33 Big Idea 4 Drill 4 295

34 Big Idea 4 Drill 4 Answers and Explanations 305

35 Free-Response Short Answers Drill 311

36 Free-Response Short Answers Drill Answers and Explanations 331

37 Free-Response Long Answers Drill 343

38 Free-Response Long Answers Drill Answers and Explanations 363

Part V: Practice Test 369

39 AP Biology Practice Test 371

40 AP Biology Practice Test Answers and Explanations 401

About the Authors 413

Part I
Using This Book
to Improve Your
AP Score

PREVIEW: YOUR KNOWLEDGE, YOUR EXPECTATIONS

Your route to a high score on the AP Biology Exam depends a lot on how you plan to use this book. Start thinking about your plan by responding to the following questions.

1. What is your level of confidence about your knowledge of the content tested by the AP Biology Exam?

    A. Very confident—I know it all
    B. I'm pretty confident, but there are topics for which I could use help
    C. Not confident—I need quite a bit of support
    D. I'm not sure

2. If you have a goal score in mind, circle your goal score for the AP Biology Exam:

    5   4   3   2   1   I'm not sure yet

3. What do you expect to learn from this book? Circle all that apply to you.

    A. A general overview of the test and what to expect
    B. Strategies for how to approach the test
    C. The content tested by this exam
    D. I'm not sure yet

YOUR GUIDE TO USING THIS BOOK

This book is organized to provide as much—or as little—support as you need, so you can use this book in whatever way will be most helpful to improving your score on the AP Biology Exam.

*   The remainder of **Part I** will provide guidance on how to use this book and help you determine your strengths and weaknesses.

*   **Part II** of this book
    o   provides information about the structure, scoring, and content of the AP Biology Exam
    o   helps you to make a study plan
    o   points you toward additional resources

- **Part III** of this book explores various strategies including
  - o   how to attack multiple-choice questions
  - o   how to write high scoring free-response answers
  - o   how to manage your time to maximize the number of points available to you

- **Part IV** of this book contains practice drills covering all of the AP Biology concepts you will find on the exams.

- **Part V** of this book contains practice tests.

You may choose to use some parts of this book over others, or you may work through the entire book. Your approach will depend on your needs and how much time you have. Let's now look at how to make this determination.

HOW TO BEGIN

1.   **Take the Diagnostic Test**

Before you can decide how to use this book, you need to take a practice test. Doing so will give you insight into your strengths and weaknesses, and the test will also help you make an effective study plan. If you're feeling test-phobic, remind yourself that a practice test is a tool for diagnosing yourself—it's not how well you do that matters but how you use information gleaned from your performance to guide your preparation.

So, before you read further, take the AP Biology Diagnostic Test starting at page 7 of this book. Be sure to do so in one sitting, following the instructions that appear before the test.

2.   **Check Your Answers**

Using the answer key that starts on page 32, count the number of questions you got right and how many you missed. Don't worry about the explanations for now, and don't worry about why you missed questions. We'll get to that soon.

3.   **Reflect on the Test**

After you take your first test, respond to the following questions::

- How much time did you spend on the multiple-choice questions?

- How much time did you spend on each essay?

- How many multiple-choice questions did you miss?

- Do you feel you had the knowledge to address the subject matter of the essays?

- Do you feel you wrote well organized, thoughtful responses to the free-response questions?

4. **Read Part II of this Book and Complete the Self-Evaluation**

As discussed on page 2, Part II will provide information on how the test is structured and scored. It will also set out areas of content that are tested.

As you read Part II, re-evaluate your answers to the questions above. At the end of Part II, you will revisit the questions above and refine your answers to them. You will then be able to make a study plan, based on your needs and time available, that will allow you to use this book most effectively.

5. **Engage with Parts III and IV as Needed**

Notice the word *engage*. You'll get more out of this book if you use it intentionally than if you read it passively, hoping for an improved score through osmosis.

Part III will open with a reminder to think about how you approach questions now and then close with a reflection section asking you to think about how and whether you will change your approach in the future.

Part IV contains drills which are designed to give you the opportunity to assess your mastery of the concepts taught in AP Biology through test-appropriate questions.

6. **Take the Practice Test, and Assess Your Performance**

Once you feel you have developed the strategies you need and gained the knowledge you lacked, you should take the practice exam at the end of this book. You should do so in one sitting, following the instructions at the beginning of the test. When you are done, check your answers to the multiple-choice sections. See if a teacher will read your essays and provide feedback.

Once you have taken the test, reflect on what areas you still need to work on, and revisit the drills in this book that address those deficiencies. Through this type of reflection and engagement, you will continue to improve.

7.   **Keep Working**

As you work through the drills, consider what additional work you need to do and how you will change your strategic approach to different parts of the test.

If you do need more guidance, there are plenty of resources available to you. Our *Cracking the AP Biology Exam* guide gives you a comprehensive review of all the biology topics you need to know for the exam and offers two practice tests. In addition, you can go to the AP Central website for more information about exam schedules and biology concepts.

AP Biology
Diagnostic Test

AP® Biology Exam

SECTION I: Multiple-Choice Questions

DO NOT OPEN THIS BOOKLET UNTIL YOU ARE TOLD TO DO SO.

At a Glance

Total Time
1 hour and 30 minutes
Number of Questions
69
Percent of Total Grade
50%
Writing Instrument
Pencil required

Instructions

Section I of this examination contains 69 multiple-choice questions. These are broken into Part A (63 multiple-choice questions) and Part B (6 grid-in questions).

Indicate all of your answers to the multiple-choice questions on the answer sheet. No credit will be given for anything written in this exam booklet, but you may use the booklet for notes or scratch work. After you have decided which of the suggested answers is best, completely fill in the corresponding oval on the answer sheet. Give only one answer to each question. If you change an answer, be sure that the previous mark is erased completely. Here is a sample question and answer.

Sample Question   Sample Answer

Chicago is a   Ⓐ ● Ⓒ Ⓓ
(A) state
(B) city
(C) country
(D) continent

Use your time effectively, working as quickly as you can without losing accuracy. Do not spend too much time on any one question. Go on to other questions and come back to the ones you have not answered if you have time. It is not expected that everyone will know the answers to all the multiple-choice questions.

About Guessing

Many candidates wonder whether or not to guess the answers to questions about which they are not certain. Multiple-choice scores are based on the number of questions answered correctly. Points are not deducted for incorrect answers, and no points are awarded for unanswered questions. Because points are not deducted for incorrect answers, you are encouraged to answer all multiple-choice questions. On any questions you do not know the answer to, you should eliminate as many choices as you can, and then select the best answer among the remaining choices.

THIS PAGE INTENTIONALLY LEFT BLANK.

BIOLOGY

SECTION I

Time—1 hour and 30 minutes

Part A: Multiple-choice Questions (63 Questions)

Directions: Each of the questions or incomplete statements below is followed by four suggested answers or completions. Select the one that is best in each case and then fill in the corresponding oval on the answer sheet.

1. Which of the following extraembryonic membranes share the same function in both reptiles and humans?

   (A) Allantois
   (B) Amnion
   (C) Chorion
   (D) Yolk sac

2. Male mallard ducks have bright green feathers on their heads and other distinctive color patterns, while female mallard ducks have brown feathers and nondistinctive markings. Which best explains this variation?

   (A) Disruptive selection
   (B) Directional selection
   (C) Stabilizing selection
   (D) Divergent evolution

3. In the operation of skeletal muscle, which of the following steps requires energy derived from ATP?

   I. Attaching myosin to actin
   II. Pulling on actin by myosin
   III. Resetting myosin to bind actin again

   (A) I
   (B) I and II
   (C) II and III
   (D) III

4. Which of the following does NOT provide evidence in support of theory of evolution?

   (A) Comparative anatomy
   (B) Embryology
   (C) Molecular biology
   (D) Mutation

5. What is a primary function of hormones?

   (A) Regulating development and behavior
   (B) Attracting potential mating partners
   (C) Running cellular respiration
   (D) Triggering action potentials in neurons

6. What is the defining characteristic of being a base?

   (A) The ability to dissolve polar substances
   (B) The release of hydroxide ions in water
   (C) A pH below 7.0
   (D) A low heat capacity

7. Labeling the CO_2 entering photosynthesis would allow a scientist to then track what being used or stored by the plant?

   (A) Chlorophyll
   (B) NADPH
   (C) Carbohydrates
   (D) ATP

8. If oxygen is not available to act as the final electron acceptor, what stage of cellular respiration would be halted first?

   (A) Electron transport chain
   (B) Glycolysis
   (C) Krebs cycle
   (D) Pyruvate dehydrogenase complex

9. When a neuron's membrane potential is between −70 millivolts and −90 millivolts, the cell is experiencing

   (A) depolarization
   (B) hyperpolarization
   (C) repolarization
   (D) threshold

GO ON TO THE NEXT PAGE.

10. Which of the following is NOT a step in transcription?

    (A) RNA polymerase binds to a promoter.
    (B) RNA polymerase creates a complementary version of both DNA strands.
    (C) RNA polymerase does not proofread the synthesized transcript.
    (D) RNA polymerase can exist with multiple copies working on multiple DNA sites.

11. What are the necessary components of a nucleotide?

    (A) An aromatic base and a five-carbon sugar
    (B) The proper ratio of carbon, hydrogen and oxygen
    (C) An aromatic base, a five-carbon sugar, and phosphate
    (D) A chain of at least three amino acids

12. How does translation convert a sequence of nucleotides into the appropriate amino acids?

    (A) A one-to-one relationship exists between nucleotides and amino acids.
    (B) Enzymes in the ribosome are responsible for converting nucleotides into amino acids.
    (C) The RNA polymerase is responsible for bringing in the correct sequence of amino acids.
    (D) Nucleotides are read in groups of three as codons to determine which amino acid is to be added to the polypeptide chain.

13. Osteoclasts are responsible for controlled destruction of bone in order to release stored calcium into the blood. Which hormone is responsible for stimulating osteoclasts?

    (A) Parathyroid hormone
    (B) Insulin
    (C) Calcitonin
    (D) ACTH

14. What characteristic best differentiates a prokaryotic cell from an animal cell?

    (A) The presence of ribosomes
    (B) The presence of a DNA genome
    (C) The presence of a cell wall
    (D) The presence of a plasma membrane

15. Meerkats live in complex social colonies where responsibilities for the group as a whole are shared amongst multiple members. Part of this work includes acting as a sentry to watch for predators or other dangers to the colony and signaling others when such dangers are detected. Given the risk inherent to the animal acting as sentry, this is an example of

    (A) territoriality
    (B) dominance
    (C) agnostic behavior
    (D) altruistic behavior

16. Which of the following is a function of the parasympathetic nervous system?

    (A) Dilation of the bronchioles
    (B) Constriction of the pupil
    (C) Limiting blood flow to the gastrointestinal system
    (D) Resolution following sexual arousal

17. Damage to the exocrine portion of the pancreas could lead to which of the following?

    I. Inability to decrease circulating glucose levels
    II. Inability to neutralize stomach acid entering the small intestine
    III. Inability to digest lipids properly

    (A) I and II
    (B) III only
    (C) II and III
    (D) I, II and III

18. Which of the following is the best example of a parasitic relationship?

    (A) Rhinoviruses residing in the upper respiratory tract of humans are the major cause of the common cold.
    (B) *E. coli*, a predominant form of bacteria in the human gut, can cause infection if accidentally moved to an open wound.
    (C) Yeast cells respiring anaerobically are killed if the concentration of ethanol they produce becomes too high in their environment.
    (D) Male lions establishing dominance in a new pride may eat existing cubs.

GO ON TO THE NEXT PAGE.

19. Two species of flowering bush exist in the same geographic area. What would provide the best evidence that the two species arose via sympatric speciation?

    (A) The internal structures of the flowers and the shapes of the leaves share common characteristics.
    (B) One of the species was imported from another country but has grown well in proximity to the other.
    (C) The same types of animals consume both species.
    (D) Sequencing of the genomes reveals similarities even though the bushes are not able to interbreed.

20. The consumption of fish, such as shark and marlin, is discouraged due to high levels of mercury present in their bodies. Why would consumption of these fish be dangerous while it remained safe to consume other types of fish normally eaten by shark and marlin?

    (A) Bioaccumulation allows mercury to build up in the larger predator fish.
    (B) The bodies of smaller fish are not impacted by mercury.
    (C) Smaller fish do not live long enough for mercury to accumulate in them.
    (D) Shark and marlin lack the ability to detoxify the mercury, whereas the bodies of other fish render it harmless.

21. Kudzu has had a negative impact as an invasive plant species because of the way it grows up around power and telephone poles. What best describes this growth pattern?

    (A) Positive gravitropism
    (B) Negative phototropism
    (C) Positive thigmotropism
    (D) Positive phototropism

22. How does the process of meiosis II differ from mitosis?

    (A) Chromosomes are separated at their centromeres.
    (B) A crossing over event can occur to increase genetic variability.
    (C) The nuclear membrane is reformed at the end of cytokinesis
    (D) Haploid cells are produced.

23. What is the genetic state of the cells produced by telophase I?

    (A) Diploid with two copies (2n2x)
    (B) Diploid with one copy (2n1x)
    (C) Haploid with two copies (1n2x)
    (D) Haploid with one copy (1n1x)

24. Which of the following will NOT impact transpiration in plants?

    (A) The amount of phloem within the plant
    (B) The presence of a cuticle on the leaves
    (C) The size of the plant's stomata
    (D) The amount of light to which the plant is exposed

25. What role does the pulmonary vein play in circulation?

    (A) It carries deoxygenated blood away from the heart.
    (B) It carries deoxygenated blood away from the organs.
    (C) It carries oxygenated blood toward the heart.
    (D) It carries oxygen to the blood.

26. What advantage does asexual reproduction tend to have in comparison to sexual reproduction?

    (A) Asexual reproduction produces genetic diversity more quickly.
    (B) Asexual reproduction typically produces new organisms at a faster rate.
    (C) Sexual reproduction provides more evolutionary pressure.
    (D) Sexual reproduction forms a larger supply of gametes.

27. Which of the following is/are reasons that breeding between members of the same species would NOT occur?

    I. Alignment of mating seasons
    II. Temporary geographic isolation
    III. Alteration of courtship behavior due to illness

    (A) I
    (B) I and II
    (C) III
    (D) II and III

GO ON TO THE NEXT PAGE.

28. How does the composition of bones in human babies differ from that in adults?

   (A) The bones of babies contain only collagen and calcium-phosphate crystals.
   (B) Cartilage is present in the bones of babies as part of developmental growth.
   (C) Androgens and estrogens promote bone growth in babies.
   (D) More calcium is stored in the bones of babies than those of adults.

29. A virus enters the lytic life cycle by incorporating its genome into that of the host cell. What is required in order for this to occur?

   (A) A dsDNA copy of the viral genome needs to be available.
   (B) Viral transcripts need to have a 5'cap.
   (C) The viral introns need to be removed and the exons spliced together into one contiguous sequence.
   (D) The enzyme reverse transcriptase needs to be made by the host cell.

30. What extraembryonic membrane is responsible for the exchange of gases and nutrients between a human fetus and its mother?

   (A) Amnion
   (B) Chorion
   (C) Placenta
   (D) Yolk sac

31. What best accounts for the diversity of cell types within a human all arising from the same genome?

   (A) Different sequences act as exons and introns creating a diversity of transcripts and producing different cellular products.
   (B) Variations in chemical exposure during fetal development impact the formation of different tissue types.
   (C) Crossing over events generate a mixture of genomes to encode differing patterns of cellular expression.
   (D) Different patterns of diet and exercise impact genome expression.

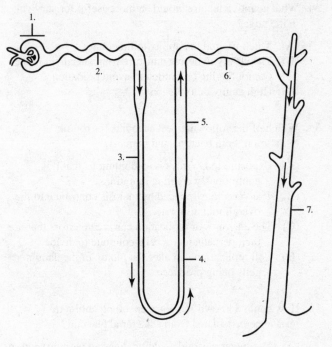

32. In which portion(s) of the nephron shown above is the tubule permeable to water at all times, assuming a gradient exists to move it?

   (A) 1 and 7
   (B) 3
   (C) 5 and 6
   (D) 4

33. What would be the best description of the placement of secondary consumers in an ecological community?

   (A) They are able to increase their population size exponentially.
   (B) They are pioneer organisms establishing a climax community.
   (C) They are always at carrying capacity for their environment.
   (D) They are k-strategists engaged in a pattern of logistic growth.

GO ON TO THE NEXT PAGE.

34. What adaptive feature would be most useful for survival in the taiga?

    (A) Thick fur and the ability to hibernate
    (B) The ability to derive moisture from succulent plants
    (C) Camouflaging coloration to avoid predation
    (D) High acuity color vision

35. Which of the following best accounts for routine mutation in both bacteria and humans?

    (A) Exposure to UV rays is so ubiquitous that it is continuously creating mutations.
    (B) The variations induced by meiosis contribute to the overall mutation rate.
    (C) The enzymes of replication can create errors that are then perpetuated as cells continue to divide.
    (D) Cell replication can alter the ploidy of the daughter cells being produced.

36. How could a known sequence encoding antibiotic resistance be isolated from a bacterial plasmid?

    (A) The intact plasmid could be digested with restriction enzymes specific to locations around the sequence.
    (B) The intact plasmid could be run on a gel electrophoresis to determine its size.
    (C) The plasmid could have an RNA copy made of its DNA sequence.
    (D) The plasmid could be transferred into another strain of bacteria to see if they also become resistant to the antibiotic.

37. How are alleles organized as part of the human genome?

    (A) Alleles are paired with their two copies proximal to one another on the same chromosome.
    (B) Alleles are paired with one copy on each member of a pair of chromosomes.
    (C) Cells have a single allele encoding a given trait.
    (D) Cells have a mix of different types of alleles on each member of a pair of chromosomes.

38. What role does the Golgi apparatus play in the production of a secretory protein?

    (A) Placement of the protein in the cell membrane
    (B) Release of the protein from the cell
    (C) Translation of the protein
    (D) Modification of the protein after translation

39. Which of the following is the best description for the function of the stomach?

    (A) Elementary digestion of carbohydrates to the level of disacharides
    (B) Elementary digestion of proteins and destruction of microorganisms
    (C) Elementary digestion of lipids for absorption by lacteals
    (D) Production of vitamin K to aid in blood clotting

40. Why does peristalsis need to slow as chyme progresses through the small intestine and enters the large intestine?

    (A) Slowed movement allows maximum exposure to the extensive surface area provided by the microvilli on the villi.
    (B) Salivary amylase is slow to digest carbohydrates and needs time in order to be effective.
    (C) Longer transit time facilitates movement of water into the gut.
    (D) Additional time is needed to neutralize the acid entering the small intestine from the stomach.

GO ON TO THE NEXT PAGE.

41. Which of the following is considered a trace element for living organisms?

    (A) Magnesium
    (B) Phosphorous
    (C) Potassium
    (D) Iron

42. Under what circumstances would apoptosis be MOST likely to occur?

    (A) A cell has experienced an error during replication.
    (B) Translation has produced more of a secretory protein than will be required.
    (C) A cell has been infected by a virus and the infection has been detected by the immune system.
    (D) The cell membrane has been breached due to physical stress.

43. Assuming that the size of a given type of finch is highly heritable, how would the mean height be impacted if a subset of smaller birds became geographically isolated for breeding purposes?

    (A) The mean height would be lower compared to the species overall.
    (B) The mean height would be higher compared to the founding population.
    (C) The mean height would be lower compared to the founding population.
    (D) The mean height would be higher compared to the species overall.

44. In a certain breed of rabbit, long narrow (N) ears are dominant over shorter rounded (n) ears. As part of a breeding experiment, one-half of the offspring have long narrow ears and the other half has shorter rounded ears. What are the likely genotypes of the parents?

    (A) NN x nn
    (B) Nn x nn
    (C) NN x Nn
    (D) Nn x nn

45. Feedback loops create regulatory mechanisms for both hormones and enzyme activity. Which of the following is NOT an example of negative feedback?

    (A) The decrease in glycolytic enzyme activity when a certain level of ATP is reached
    (B) High estrogen and progesterone levels preventing the release of FSH and LH
    (C) Decreasing acid production in the stomach when food is not present
    (D) The continued production of prolactin based on frequent nursing patterns of a newborn

46. Which of the following is a requirement to maintain allelic frequencies in a Hardy-Weinberg population model?

    (A) Genetic drift
    (B) Natural selection
    (C) Random mating
    (D) Frequent mutation

47. Which of the following is NOT a characteristic of viruses?

    (A) The ability to infect bacterial cells
    (B) Genomes that can be composed of DNA or RNA
    (C) The use of plasmids to increase genetic diversity
    (D) Genomes that mutate quickly

48. What are some identifying structural components of an amino acid?

    (A) A phosphate and a ribose
    (B) An equal ratio of carbon to hydrogen
    (C) A peptide bond and a disulfide bridge
    (D) A variable group and an amino group

GO ON TO THE NEXT PAGE.

49. Which of the following is the best description of the events in mitosis?

    (A) A cell replicates its genome, divides the information into two matching copies, and splits the contents of the cell to produce two daughter cells.

    (B) A cell divides its diploid genome into two equivalent copies and splits the contents of the cell to produce two daughter cells.

    (C) A cell replicates its genome, divides the information into two matching copies, and provides one daughter cell with the majority of its cellular contents.

    (D) A cell provides the opportunity for chromosomes to recombine and then replicates prior to dividing its cellular contents between daughter cells.

50. How does a chemoautotroph function metabolically?

    (A) CO_2 as its carbon source, sunlight as its energy source

    (B) Sunlight as its carbon source, organic molecules as its energy source

    (C) Organic molecules as its carbon and energy source

    (D) CO_2 as its carbon source, inorganic molecules as its energy source

51. In eukaryotes, the electron transport chain occurs across which membrane?

    (A) Plasma membrane
    (B) Inner mitochondrial membrane
    (C) Endoplasmic reticulum
    (D) Nuclear membrane

52. What types of membrane channels are used to maintain resting membrane potential?

    (A) K^+ leak channels and Na^+ leak channels
    (B) Na^+/K^+ pump and K^+ leak channels
    (C) Na^+/K^+ pump and Na^+ leak channels
    (D) Ca^{++} leak channels and Cl^- leak channels

GO ON TO THE NEXT PAGE.

Questions 53-55 refer to the diagram.

The pentose phosphate pathway is one of several metabolic pathways that complement the central processes producing ATP for the body.

Pentose Phosphate pathway

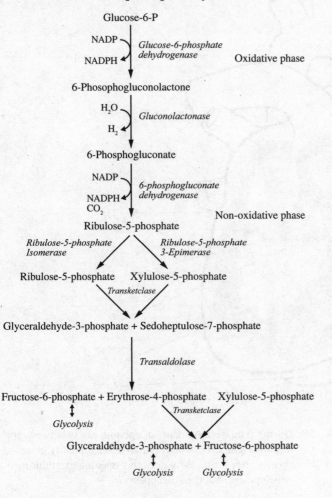

53. How does the pentose phosphate pathway contribute to the production of nucleotides?

    (A) Producing ribose-5-phosphate
    (B) Producing NADPH
    (C) Producing ribulose-5-phosphate
    (D) Producing fructose-6-phosphate

54. In what way does the pentose phosphate pathway interact in a circular fashion with glycolysis?

    (A) High levels of glyceraldehyde-3-phosphate downregulate the pathway and leave glucose-6-phosphate in glycolysis.
    (B) High levels of glucose-6-phosphate power both the pathway and glycolysis.
    (C) Glucose-6-phosphate enters the pathway from glycolysis, and fructose-6-phosphate leaves the pathway for glycolysis.
    (D) $NADP^+$ enters the pathway from glycolysis, and NADPH leaves the pathway for glycolysis.

55. What would be the most effective target for negative feedback in order to regulate the pentose phosphate pathway?

    (A) 6-phosphogluconate dehydrogenase
    (B) Transketolase
    (C) Transaldolase
    (D) Glucose-6-phosphate dehydrogenase

GO ON TO THE NEXT PAGE.

Questions 56-58 refer to the diagram.

The heart separates the circulatory system into two divisions: the pulmonic and the systemic. The flow of blood through the heart is determined by where it originates and whether or not it is oxygenated.

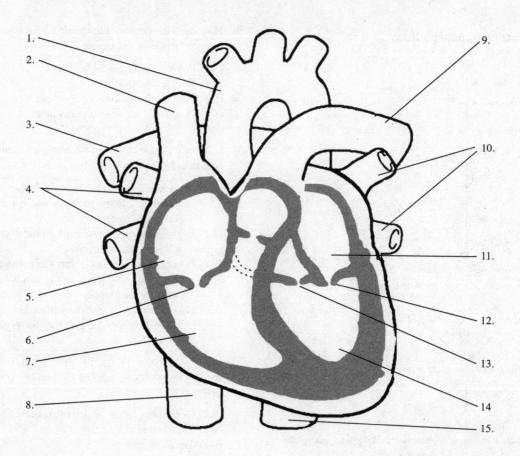

56. Which of the following is an accurate sequence for the movement of oxygenated blood through the heart?

   (A) 2 – 6 – 7 – 9
   (B) 10 – 12 – 14 – 1
   (C) 8 – 4 – 5 – 9
   (D) 3 – 14 – 11 – 10

57. What valves are present at locations 6 and 12, respectively?

   (A) A tricuspid semilunar valve and a bicuspid semilunar valve
   (B) A bicuspid AV valve and a tricuspid AV valve
   (C) A tricuspid AV valve and a bicuspid AV valve
   (D) A bicuspid semilunar valve and a tricuspid semilunar valve

58. During fetal development, an opening exists between the left and right atria. The foramen ovale typically closes shortly after birth. Why does this opening exist during development, but then disappears?

   (A) The foramen ovale allows the fetus to more effectively distribute the oxygen supplied by the mother's body and thus return any extra to maternal circulation.
   (B) The foramen ovale allows for the mixing of blood between the circuits of the body because the lungs are not yet in use.
   (C) The foramen ovale exists as a vestigial structure which is then corrected once fetal development is complete.
   (D) The foramen ovale allows space for the heart muscle to grow during fetal development so the chambers can form properly.

GO ON TO THE NEXT PAGE.

Questions 59-61 refer to the diagram.

The menstrual cycle is controlled by the anterior pituitary regulating the ovaries, the ovaries regulating the uterus and the uterus feeding back to the anterior pituitary. This graph shows the relationship between the relevant hormones over the course of a typical 28-day cycle.

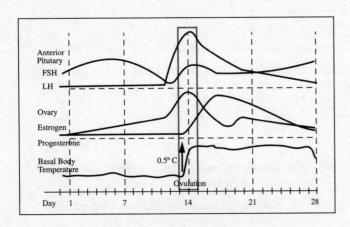

59. In the latter half of the menstrual cycle, basal body temperature (BBT) undergoes a slight but detectable and sustained rise. Based on the graph, which hormone is most likely responsible for this rise?

    (A) Progesterone
    (B) LH
    (C) FSH
    (D) Estrogen

60. Given that FSH and LH levels fall during the latter half of the menstrual cycle, what impact are estrogen and progesterone together likely having on the anterior pituitary?

    (A) Creating a positive feedback loop
    (B) Creating a process of feedforward regulation
    (C) Creating allosteric regulation
    (D) Creating a negative feedback loop

61. During what days is the corpus luteum an active structure on the ovary?

    (A) 1-5
    (B) 5-14
    (C) 14-15
    (D) 16-24

GO ON TO THE NEXT PAGE.

Questions 62-63 refer to the diagram.

The diagram depicts how a sequence originating in the genome then become interpreted by transcription and finally by translation.

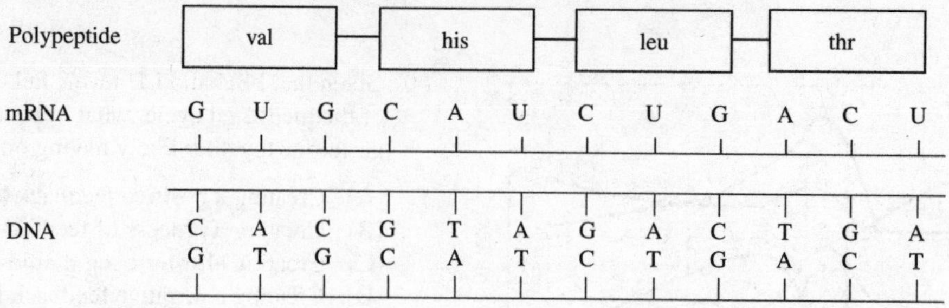

| Polypeptide | val | his | leu | thr |

| mRNA | G | U | G | C | A | U | C | U | G | A | C | U |

| DNA | C | A | C | G | T | A | G | A | C | T | G | A |
| | G | T | G | C | A | T | C | T | G | A | C | T |

62. If the codon for histidine was removed, what type of mutation would have been achieved?

(A) Frameshift
(B) Deletion
(C) Transition
(D) Insertion

63. What best describes the relationship between the DNA sequence and the transcript derived from it?

(A) The two sequences are identical, except for the exchange of thymine for uracil
(B) The two sequences exist in a 3:1 ratio of nucleotides.
(C) The two sequences are complementary.
(D) The two sequences could both undergo translation.

GO ON TO THE NEXT PAGE.

<u>Directions:</u> This part B consists of questions requiring numeric answers. Calculate the correct answer for each question.

64. In a population model to study squirrels, the frequency of a dominant allele for tail length is found to be 0.8. What is the frequency of squirrels who are heterozygous for this allele?

66. A man with the blood genotype I^AI^B has children with a woman with the blood genotype I^Ai. What is the probability that of their two children one child will have blood type A and one child will have blood type B?

65. The average male African bush elephant weighs approximately 12,000 lbs and has a trophic level efficiency of about 5%. If 650 tons of plants are available in a wildlife preserve for consumption by elephants, how many tons of them can be maintained in the preserve?

GO ON TO THE NEXT PAGE.

Question 67 refers to the diagram.

The human eye contains three types of cones which each absorb light from a different part of the visible spectrum. The input from all three types creates high acuity color vision that is integrated by the brain.

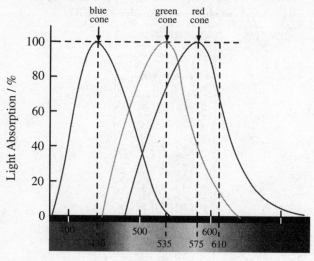

Question 68 refers to the diagram.

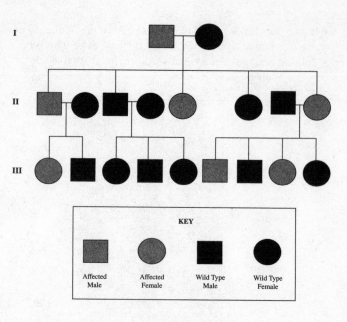

KEY

Affected Male	Affected Female	Wild Type Male	Wild Type Female

68. If an affected person from this family pedigree has a child with someone who is not affected, what is the probability that that child will have the condition?

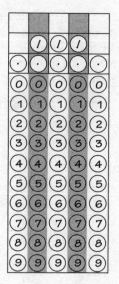

67. At what wavelength are both red and blue cones absorbing approximately 40% of the available light?

GO ON TO THE NEXT PAGE.

<u>Question 69</u> refers to the diagram.

Bacteria are able to maintain log growth so long as nutrients are in continual supply and metabolic waste products are removed from the environments. Often these conditions can only be achieved via an *in vitro* environment.

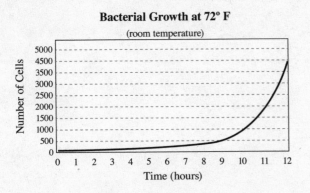

Bacterial Growth at 72° F
(room temperature)

69. Given the growth demonstrated in the graph, approximately how many bacterial cells will be present in a laboratory culture after 13 hours?

END OF SECTION I

This page intentionally left blank.

BIOLOGY
SECTION II

Planning time—10 minutes

Writing time—1 hour and 30 minutes

<u>Directions:</u> Questions 1 and 2 are long-form essay questions that should require about 20 minutes each to answer. Questions 3 through 8 are short free-response questions that should require about 6 minutes each to answer. Read each question carefully and write your response. Answers must be written out. Outline form is not acceptable. It is important that you read each question completely before you begin to write.

1. The kidney is a highly vascularized organ composed of both nephrons and their surrounding network of blood vessels.

    a. Describe the stages of the nephron, including their major function and any associated blood vessels.

    b. Define what components of filtrate end up in the urine to be excreted and why the presence of white blood cells is cause for concern.

GO ON TO THE NEXT PAGE.

2. The polymerase chain reaction (PCR) is an experimental technique used to create abundant copies of an existing DNA sequence.

   a. Explain the process by which PCR is able to produce multiple DNA copies.

   b. Describe an experiment in which PCR could be used to locate a gene of interest.

   c. Define the unique properties of Taq polymerase.

GO ON TO THE NEXT PAGE.

3. Define natural selection, and give two examples.

4. Describe the process of glycolysis. Explain the differences between it occurring aerobically versus anaerobically.

GO ON TO THE NEXT PAGE.

5. Describe the two possible life cycles of a bacteriophage.

6. Define the functions of the sympathetic nervous system, and give an example of its type of response.

GO ON TO THE NEXT PAGE.

7. Describe the process of depolarization and repolarization in a neuron.

8. Define the impact of blood pH on respiratory rate.

STOP

END OF EXAM

AP Biology
Diagnostic Test
Answers and
Explanations

AP BIOLOGY DIAGNOSTIC TEST ANSWER KEY

1.	B	38.	D	
2.	A	39.	B	
3.	D	40.	A	
4.	D	41.	D	
5.	A	42.	C	
6.	B	43.	A	
7.	C	44.	B	
8.	A	45.	D	
9.	B	46.	C	
10.	B	47.	C	
11.	C	48.	D	
12.	D	49.	A	
13.	A	50.	D	
14.	C	51.	B	
15.	D	52.	B	
16.	B	53.	A	
17.	C	54.	C	
18.	A	55.	D	
19.	D	56.	B	
20.	A	57.	C	
21.	C	58.	B	
22.	D	59.	A	
23.	C	60.	D	
24.	A	61.	D	
25.	C	62.	B	
26.	B	63.	C	
27.	D	64	0.32	
28.	B	65.	5	
29.	A	66.	$\frac{1}{4}$	
30.	C			
31.	A	67.	500	
32.	B	68.	$\frac{1}{2}$	
33.	D			
34.	A	69.	9000	
35.	C			
36.	A			
37.	B			

ANSWERS AND EXPLANATIONS

Section I

1. **B** The amnion is a fluid-filled sac that protects both a reptile fetus in its shell as well as a human fetus in the womb. While both reptiles and humans have the other three extraembryonic membranes, they serve different functions in each.

2. **A** Disruptive selection leads to extreme traits within a member of the same species, thus differentiating one subset of that species from another. Common traits between the subsets are selected against.

3. **D** The only step in the sliding filament process that uses energy from ATP is the resetting of the myosin heads to prepare them to bind to actin again. Detaching the myosin heads from actin requires ATP to bind, but it is not yet hydrolyzed.

4. **D** Mutation contributes to genetic diversity, but it is not a main source of evidence concerning the process of evolution. The similarities found throughout comparative anatomy, embryology, and molecular biology all do provide that scientific support.

5. **A** Hormones are involved in our behavior, physical development, growth, and reproductive cycles. Pheromones contribute to attraction, enzymes run cellular respiration and neurotransmitters trigger action potentials.

6. **B** The defining characteristic of bases is their release of hydroxide ions when mixed in solution, thus contributing to an alkaline state and pH level above 7.0.

7. **C** The CO_2, which is a necessary part of photosynthesis, becomes incorporated into the carbohydrate ($C_6H_{12}O_6$) molecules being made and stored by the plant for later use.

8. **A** The electron transport chain is dependent upon oxygen to act as the final electron acceptor so that the high energy electron carriers (NADH, $FADH_2$) can be regenerated.

9. **B** Hyperpolarization describes a cell whose membrane potential is below resting membrane potential, which is –70 millivolts.

10. **B** RNA polymerase is responsible for synthesizing transcripts which are single-stranded and thus represent a complementary RNA-version of only one DNA strand.

11. **C** Nucleotides are composed of a five-carbon sugar (ribose in RNA, deoxyribose in DNA), an aromatic base and phosphate group. Without the phosphate group, the molecule is a nucleoside.

12. **D** Codons refer to the grouping of three nucleotides in a transcript which then determine the amino acid to be added during the process of translation. Those codons are read by the anticodon loop on the tRNA which then delivers the needed amino acid.

13. **A** Parathyroid hormone increases the body's circulating levels of calcium by stimulating osteoclasts to break down bone. This regulated phagocytic activity is part of the constant bone remodeling that goes on even after full adult height is achieved.

14. **C** Animal cells do not have a cell wall; the other three components are present in both prokaryotic and animal cells. Remember prokaryotes do not lack all organelles; rather, they do not have membrane-bound organelles.

15. **D** Altruistic behavior serves the good of the group, even if the individual is put at risk. This does not involve the protection of land, attempts to dominant other members of the group, or aggressive behavior in an attempt to protect resources.

16. **B** The parasympathetic nervous system is responsible for near vision, which requires contraction of the pupil so the eye can focus on what is close. Bronchodilation, the slowing of digestion, and resolution after sexual arousal are all sympathetic nervous system responses.

17. **C** The endocrine components of the pancreas provide a source of very alkaline aqueous buffer that is released to the small intestine to neutralize stomach acid as well as supply several digestive enzymes, including the lipases that allow the body to break down and absorb lipids. Insulin allows the body to regulate circulating levels of glucose, but it is produce by the endocrine portions of the pancreas.

18. **A** Rhinoviruses are causing harm to their host in order to reproduce and be spread; in a parasitic relationship, there needs to be dependence between an organism (the parasite) and its host in which the host is harmed in the course of the interaction.

19. **D** Sympatric speciation occurs when separate species arise, even in the absence of physical separation. The lack of an ability to interbreed is a primary characteristic of creating distinct species, yet genetic similarities would indicate a common evolutionary ancestor. Structural similarities are not as compelling.

20. **A** Shark and marlin are either secondary or tertiary consumers in their communities. Due to the amount of biomass they consume, a large volume of mercury enters their bodies and accumulates there. Smaller fish consume less biomass and thus have less opportunity for mercury to accumulate at toxic levels.

21. **C** Positive thigmotropism describes a plant's response to touch, and the growth pattern described here involves contact with other structures. The kudzu's growth is not being dictated in this case by available light or growth in accord with gravity.

22. **D** The end product of meiosis II is the production of four haploid gametes. Chromosomes separating at the centromere and the reformation of the nuclear membrane occur in both and crossing over only occurs in meiosis I.

23. **C** At the end of telophase I in meiosis I, the cell division produces two cells that are both haploid, and each have two copies of that half of the genome. Replicated cells are 2n2x, and cells at the end of telophase II are 1n1x.

24. **A** Transpiration describes the movement of water through a plant, including evaporation of water from its surfaces. Phloem is the tissue moving nutrients away from the leaves and does not move water.

25. **C** Veins carry blood toward the heart and the pulmonary vein is the vessel that returns from the lungs carrying oxygenated blood, which will then be pumped by the heart out into the systemic circulation in order for that oxygen to be used by the body.

26. **B** The primary advantage to asexual reproduction is the speed at which new organisms are created. It does not generate genetic diversity or gametes and any relationship between sexual reproduction, and evolution is not addressing a benefit of asexual reproduction.

27. **D** Physical separation and changes in behavior related to mating would both be barriers to breeding. Having the mating seasons align properly would actually promote breeding within the species.

28. **B** Cartilage is deposited at the growth plate to extend the length of bone during early growth so it is part of the skeleton early in life. By adulthood, the bones are all collagen and calcium-phosphate crystals, and growth is actually halted by extended exposure to sex hormones.

29. **A** Host cell genomes are composed of dsDNA, so the viral genome must match this format in order to be incorporated. This process does not involve transcription and reverse transcriptase. If transcriptase is needed by an ssRNA virus, would be synthesized based on the viral genome and would not be native to the host cell.

30. **C** The placenta is responsible for the transfer of material between mother and fetus in humans. The chorion is the layer of this membrane derived from the fetus directly, but both the maternal and fetal components are necessary for proper functioning.

31. **A** All the cells within a given person possess the same genome, but different cells utilize different sequences within that genome to encode needed products. This process occurs throughout life and not just during fetal development. Crossing over is restricted to gamete formation, and lifestyle patterns are not sufficient to account for these differences.

32. **B** The descending loop of Henle is permeable only to water and not ions at all times. Movement of water out of the tubule is based on current concentrations but is always possible.

33. **D** Due to the volume of biomass they require, secondary consumers need to be k-strategists who exist within the limitations of logistic growth. They are not in an ecological position to found communities or grow exponentially.

34. **A** The taiga is characterized by long, cold winters, so the ability to survive in such atmospheric conditions become key. Succulent plants do not grow in the taiga and although camouflage and high acuity color vision are generally useful, they do not address the specific conditions of the taiga.

35. **C** Replicative enzymes do proofread the sequence being constructed, but they are not perfect. UV exposure could not account for all routine mutation (A), bacteria do engage in meiosis (B), and replication does not alter ploidy because the daughter cells are copies of the parent cell (D).

36. **A** Restriction enzymes digest DNA at sites with specific short sequences. If the sequence of the plasmid is known, then these could be used to extract the sequence of interest.

37. **B** Humans are diploid, so there are pairs of chromosomes and paired alleles exist in matched positions on each chromosome. There can be many variations on a given allele within a population, but a particular person will either have two copies that are the same (being homozygous) or two copies that happen to be different (heterozygous).

38. **D** Secretory proteins are released from the cell by vesicles, but they need to undergo post-translational modification in the Golgi apparatus before being packaged for release.

39. **B** Stomach acid destroys microorganisms and activates enzymes to begin digesting proteins. Salivary amylase starts breaking down carbohydrates in the mouth, lipids do not begin to be digested until reaching the small intestine, and vitamin K is made by the bacteria in the large intestine.

40. **A** The villi and microvilli are responsible for absorbing nutrients in their monomer form and the more time they are exposed to the digested food, the more of that which can be absorbed. While the acid leaving the stomach does need to be neutralized, this is done almost immediately by high pH buffer coming from the pancreas to avoid damaging the small intestine.

41. **D** Iron is considered a trace element because of the minimal amount that is required. The others are needed less than carbon or oxygen but are still more present than iron.

42. **C** Apoptosis is a controlled form of cell death that is used to shut down cells which are no longer useful or somehow present a threat, as a virally infected cell would. A cell with a ruptured membrane would already be dead.

43. **A** Because the smaller birds have been isolated and because height is very heritable, the overall height of this subpopulation will trend downwards, and therefore the mean would be lower when compared to the species overall.

44. **B** In classical dominance a 50/50 split between the dominant and recessive trait in the offspring is triggered by the parent with the dominant phenotype being a heterozygote and the parent with the recessive phenotype being homozygous.

45. **D** The nursing newborn drains milk from the breast which signals the further production of prolactin. This cycle continues so long as the breast is being emptied and thus represents positive feedback rather than a negative feedback loop.

46. **C** The Hardy-Weinberg model of population genetics depends on maintaining consistent allelic frequencies; thus all mating would need to be entirely random as opposed to mating that would favor certain traits, and thus potentially make them more predominant in the population.

47. **C** Plasmids are extrachromosomal genetic elements found in bacterial cells. They are not part of the genetic composition of a virus.

48. **D** Amino acids all contain an amino group as part of their baseline structure. The specific identity of an amino acid is determined by which variable group, or R group, is present. The 20 different variables groups form the 20 different amino acids.

49. **A** The goal of mitosis is to produce two daughter cells that are genetically and structurally equivalent to the parent cell. This is achieved by replicating the genome and ensuring equal division of the cellular contents without any changes to the genetic configuration.

50. **D** A chemoautotroph uses CO_2 to provide carbon and inorganic molecules to provide energy.

51. **B** The electron transport chain utilizes the inner mitochondrial membrane in eukaryotes; in prokaryotes, it utilizes the plasma membrane, the only membrane that type of cell has.

52. **B** Resting membrane potential is maintained as a net negative on the interior of the cell by the Na^+/K^+ pump removing more positively charge ions than it lets in and by the K^+ leak channels.

53. **A** Ribose-5-phosphate is the pentose phosphate pathway intermediate that is a subcomponent of nucleotides.

54. **C** The circular nature of the process needs to be reflected in how each can supply material to the other in something of a mutual exchange. Glucose-6-phosphate is made by glycolysis and used in the pathway while fructose-6-phoshphate is also made by the pathway and can be used in glycolysis.

55. **D** For negative feedback to be most effective, it needs to stop a process as early as possible to avoid wasted production of intermediates if the pathway is not currently needed. Thus, the earliest enzymes in the process would be the best to target for regulation.

56. **B** Oxygenated blood returns to the heart via the pulmonary veins, enters the left atrium, gets pumped to the left ventricle, and leaves the heart for the systemic circulation via the aorta.

57. **C** The valves between the atria and the ventricles are referred to as AV (atrioventricular) valves. The one on the right side of the heart has three flaps (tricuspid) while the one on the left has two (bicuspid).

58. **B** During fetal development, the lungs are not being used to supply oxygen to the body; rather, all of the fetus' oxygen is derived from the maternal blood supply, and thus the pulmonic circuit is not needed. This need quickly changes after birth once the baby is dependent on air and breathing to supply oxygen, and thus the pulmonic circuit needs to be fully separated from the systemic circuit.

59. **A** The hormone which rises in the latter half of the menstrual cycle and stays elevated, like basal body temperature, is progesterone and the most likely cause of the change.

60. **D** The rise in estrogen and progesterone suppresses production of FSH and LH by the anterior pituitary, which is an example of negative feedback establishing a pattern of hormone regulation.

61. **D** The corpus luteum is the hormone secreting structure left on the ovary after ovulation and thus is active in the latter half of the menstrual cycle.

62. **B** The removal of an entire codon does remove an amino acid, but it does not alter the reading frame and thus the mutation is a deletion.

63. **C** mRNA is complementary to its DNA template. There is a 1:1 relationship between DNA and RNA, and only RNA can undergo translation.

64. **0.32** In modeling population genetics, p represents the dominant allele, q represents the recessive allele, and $p + q = 1$. For this problem, $p = 0.8$ so $q = 0.2$. The heterozygous frequency is represented by $2pq$ which makes the result 0.32.

65. **5** First, convert the elephant weight to tons because the plants are given in tons: 12,000 lbs = 6 tons. Next, do the conversion of trophic efficiency for one elephant: 6 tons/ plant tons = 5 /100. This indicates that 120 tons of plants are needed to support one elephant. Finally, divide 120 into 650; the answer is 5 with a remainder of 50, but that remainder is not enough to support an additional elephant.

66. $\dfrac{1}{4}$ The probability that the couple has a child with blood type A is $\dfrac{1}{2}$ (the child will be either $I^A I^A$ or $I^A i$), and the probability that the couple has a child with blood type AB is also $\dfrac{1}{2}$. Because the question asks for the probability of a type A child AND a type AB child, the two probabilities must be multiplied giving $\dfrac{1}{4}$.

67. **500** The *y*-axis gives the percent of light being absorbed so start at 40 and trace over to both the blue and red cone curves. Because the question asks about it being the same for both, look for where the graphs overlap. The approximate 40% absorption for both is occurring at 500 nm.

68. $\frac{1}{2}$ The condition must be autosomal dominant: no generations are skipped, and there is a relatively equal distribution between affected males and females in addition to the fact that one parent in Generation I is passing down the condition. However, the father in Generation I must be heterozygous because only half of his children are affected; if he was homozygous all would be affected. Thus, the affected offspring in Generations II and III must also be heterozygous. If any of them has a child with an unaffected individual, the chance of that child being affected is $\frac{1}{2}$.

69. **9,000** The graph showing the bacterial population in the lab doubling every hour. At 12 hours there are approximately 4,500 bacterial cells, so at 13 hours there would be 9,000.

Section II

1. a. The kidney is made up of about a million tiny structures known as nephrons. Each nephron consists of the Bowman's capsule, the proximal convoluted tubule, the loop of Henkle, the distal convoluted tubule, and the collecting duct. The Bowman's capsule has a ball of capillaries inside it called a glomerulus, where the blood is filtered. Filtered items include ions, water, glucose, urea and amino acids. The filtrate then leave the capsule and passes through the proximal convoluted tubule, where some of the larger materials are reabsorbed. The descending part of the loop of Henle is where water in reabsorbed into the capillaries. The ascending part of the loop of Henle reabsorbs ions into the capillaries. The filtrate then passes through the distal convoluted tubule. If aldosterone is present, then sodium ions will be reabsorbed here. Finally the urine will proceed to the collecting duct. If vasopressin (also known as antidiuretic hormone or ADH) is present in the system, then water will be reabsorbed and the urine will be more concentrated.

   b. Urine typically consists of water, urea and ions. Urea is a nitrogenous waste product that the body synthesizes from amino acid oxidation or ammonia. Typical ions found in urine include sodium, potassium, and chloride ions. The rate at which each filtrate appears is based on the needs of the body. For example, if a person is dehydrated then there will most likely be less water in the urine. Typically the glomerulus only allows for the filtration of small chemicals; as such most cells, lipid, and proteins do not appear in the urine. If there are white blood cells in the urine, this is an indication of an infection in the kidney or urinary system.

2. a. The polymerase chain reaction (otherwise known as PCR) produces multiple DNA copies through a multi-step process of heating and cooling. First a PCR tube is prepared that contains DNA, primers, *Taq* polymerase, nucleotides and a buffer solution. First the PCR tube mixture is heated to break the hydrogen bonds and separate DNA strands. Then the mixture is cooled to allow the primers to anneal to the DNA sequence that the scientist wants to copy. Once that is complete, the mixture is warmed again to allow the *Taq* polymerase to bind to the primers and add nucleotides on each end of the DNA strand. Once the first cycle is finished, two identical double-stranded DNA molecules are created.

    b. There are many different kinds of experiments that a scientist can use PCR for. PCR can be used to evaluate genomes between different species of bacteria to determine the presence of similar genes. It can also be used to evaluate the relatedness of two people for inclusion into a family pedigree. Another experiment would be to evaluate genes in different tumors to study treatment options.

    c. *Taq* polymerase is unique because of its heat-resistant properties. Most enzymes are proteins which are typically denatured when they are placed in high temperature solutions. *Taq* polymerase is an enzyme that remains intact in high temperatures and is also functional at lower temperatures to aid in DNA synthesis.

3. Natural selection refers to a phenomenon where certain organisms become more "fit" for an environment because they possess certain biological traits that are beneficial to their survival. The organisms that are the fittest to survive are more likely produce a second generation of offspring. Note that natural selection is the opposite of Lamarck's theory of evolution which stated that acquired traits are passed on from parent to offspring. There are may examples that you can use for natural selection. One example is the fact that salamanders can change color and use camouflage to evade predators. At some point in the past, there were salamanders that could or could not change color. However, the color-changing salamanders were able to avoid being eaten and continued to reproduce while the non-color changing salamanders eventually died out. Another example are seals. Seals that are able to store more brown fat were more able to survive in colder temperatures that seals that did not have the ability.

4. Glycolysis is the process by which glucose is broken down into two pyruvic acid molecules, creating ATP (adenosine triphosphate, a molecule packed with enormous amounts of energy) in the process. Glycolysis occurs in the cytoplasm and during this process, glucose becomes phosphorylated and isomerized. In sum, 2 ATP are used by the cells to break down the glucose for the creation of 4 ATP molecules and two NADH molecules. Glycolysis can be an aerobic or an anaerobic process. In aerobic conditions, the glucose is fully oxidized as the carbons move through the Krebs cycle and oxidative phosphorylation. In anaerobic conditions, an alternate final electron acceptor is necessary because of the absence of oxygen. Anaerobic organisms undergo fermentation after glycolysis, with the final products being lactic acid or ethanol. Another difference is that anaerobic respiration after glycolysis only produces a net gain of 2 ATP as opposed to a net gain of 36 ATP in aerobic respiration after glycolysis.

5. Bacteriophages undergo two life cycles: the lytic cycle and the lysogenic cycle. In the lytic cycle, the genetic code of a virus enters a bacterium and starts using the host's machinery to replicate the virus's genetic material. The virus's genome also directs the host to create protein capsids so that the cell can create new viruses. The virus will then cause the cell to lyse (burst open) releasing the new viruses to go on and infect other hosts. In the lysogenic phase, the virus invades a host cell and inserts its genome in the host cell's genome but does not immediately replicate. Instead, the virus lays dormant. Any daughter cells that the host cell produces will contain the virus genome. Eventually an environmental or stress trigger will cause the virus to enter into the lytic phase.

6. The sympathetic nervous system is known as the system that controls the body's "fight-or-flight" response and allows the body to handle stress (either by avoiding danger or by defending against an attack). This system works by raising the heart rate to increase blood supply to the skeletal muscles. Alternately, this system can increase the respiratory rate to allow oxygen to be quickly delivered to body tissues. Once the threat has subsided the parasympathetic nervous system will bring the body back to homeostasis. A classic example would be a person who is in a burning building. The sympathetic nervous system would kick in, allowing the person to move quickly to escape through any nearby exits and, if need be, be able to quickly pry a door open that a person would not normally be strong enough to open.

7. Neurons undergo depolarization and repolarization in order to transmit impulses and information to other neurons or organs. When depolarization occurs, the cell potential rises from a resting rate of –70 mV to a threshold of –50 mV due to the entrance of sodium (Na^+) ions. Once the cell reaches threshold, the cell fires and triggers voltage-gated sodium ion channels to open which switches the polarity form a negative to a positive charge of +35 mV. Repolarization occurs once the sodium ions have flooded the neuron, by which point the sodium channels close. At this point, potassium (K^+) channels open and potassium ions rush out of the cell. The electrical charges reverse again because the neuron becomes negatively charged. Note that while the cell is more negative again, the sodium ions are all located inside the cell and the potassium ions are outside of the cell and this must be reset. This ion distribution is reestablished by using the sodium-potassium pump to move the Na^+ and K^+ ions to its proper locations.

8. In general, blood pH is meant to be slightly alkaline (basic), with an average pH of 7.4. When an organism is breathing too slowly, CO_2 will accumulate and cause the blood pH to drop. This will cause the organism to breathe more to restore the pH balance. Conversely, when a body is breathing too fast, too much CO_2 is lost. The blood pH will rise and drive the body to breathe less to bring the blood back to homeostasis.

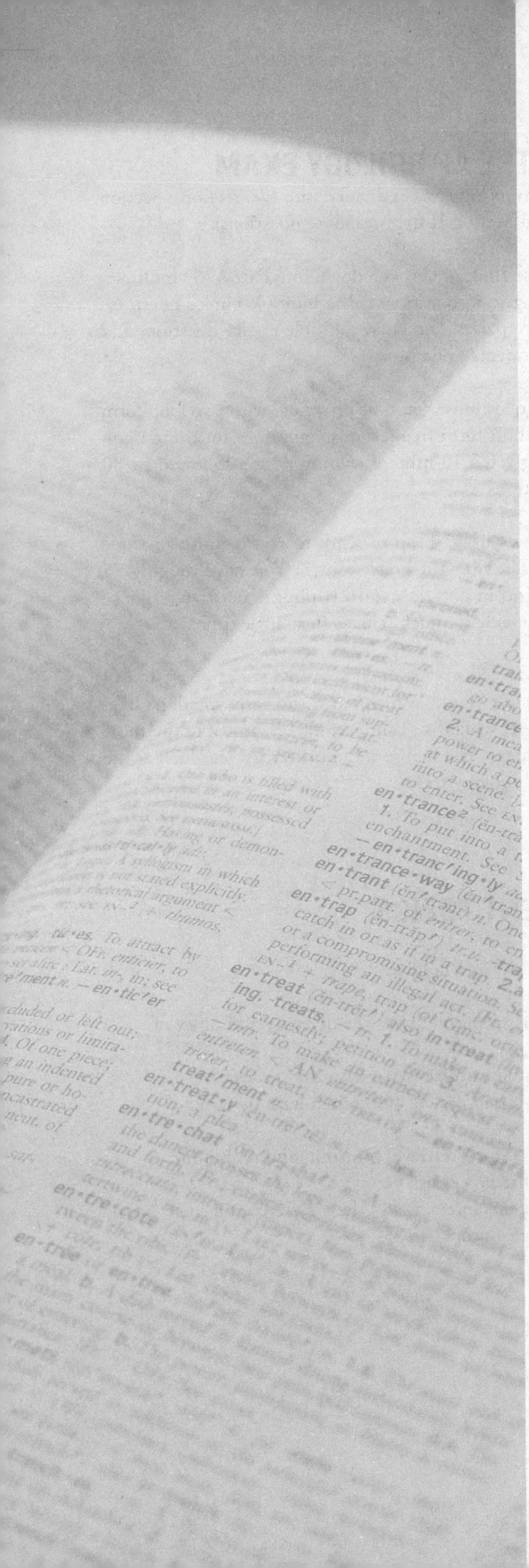

Part II
About the AP Biology Exam

THE STRUCTURE OF THE AP BIOLOGY EXAM

The AP Biology Exam is three hours long and is divided into two sections: Section I (multiple-choice questions) and Section II (free-response questions).

Section I consists of 69 questions. These are broken down into Part A (63 multiple-choice questions) and Part B (6 grid-in questions). The multiple-choice questions are further broken down into two parts: (1) regular multiple-choice questions and (2) questions dealing with experiments or data.

Section II involves free-response questions. You'll be presented with two long-form free-response questions and six short-form free-response questions touching upon key issues in biology. You'll be given a 10-minute reading period followed by 80 minutes to answer all eight questions.

If you're thinking that this sounds like a heap of work to try to finish in three hours, you're absolutely right. Here's how it breaks down: You have roughly 75 seconds per multiple-choice or grid-in question and 21 minutes per free-response question. How can you possibly tackle so much science in so little time?

Fortunately, there's absolutely no need to. As you'll soon see, we're going to ask you to leave a small chunk of the test blank. Which part? The parts you don't like. This selective approach to the test, which we call "pacing," is probably the most important part of our overall strategy. But before we talk strategy, let's look at the topics that are covered by the AP Biology Exam.

OVERVIEW OF CONTENT TOPICS

The AP Biology Exam covers these four Big Ideas.

- Big Idea 1: The process of evolution drives the diversity and unity of life.
- Big Idea 2: Biological systems utilize free energy and molecular building blocks to grow, to reproduce, and to maintain dynamic homeostasis.
- Big Idea 3: Living systems store, retrieve, transmit, and respond to information essential to life processes.
- Big Idea 4: Biological systems interact, and these systems and their interactions possess complex properties.

These four areas are further subdivided into major topics. These topics include the following:

1. Chemistry of Life
   - Organic molecules in organisms
   - Water
   - Free-energy changes
   - Enzymes
2. Cells
   - Prokaryotic and eukaryotic cells
   - Membranes
   - Subcellular organization
   - Cell cycle and its regulation
3. Cellular Energetics
   - Coupled reactions
   - Fermentation and cellular respiration
   - Photosynthesis
4. Heredity
   - Meiosis and gametogenesis
   - Eukaryotic chromosomes
   - Inheritance patterns
5. Molecular Genetics
   - RNA and DNA structure and function
   - Gene regulation
   - Mutation
   - Viral structure and replication
   - Nucleic acid technology and applications
6. Evolutionary Biology
   - Early evolution of life
   - Evidence for evolution
   - Mechanism of evolution
7. Diversity of Organisms
   - Evolutionary patterns
   - Survey of the diversity of life
   - Phylogenetic classification
   - Evolutionary relationships
8. Structure and Function of Plants and Animals
   - Reproduction, growth, and development
   - Structural, physiological, and behavioral adaptation
   - Response to the environment
9. Ecology
   - Population dynamics
   - Communities and ecosystems
   - Global issues

This might seem like an awful lot of information. But for each topic, there are just a few key facts you'll need to know. Your biology textbooks may go into far greater detail about some of these topics than you actually need to know for the exam.

HOW AP EXAMS ARE USED

Different colleges use AP Exams in different ways, so it is important that you go to a particular college's web site to determine how it uses AP Exams. The three items below represent the main ways in which AP Exam scores can be used.

- **College Credit.** Some colleges will give you college credit if you score well on an AP Exam. These credits count toward your graduation requirements, meaning that you can take fewer courses while in college. Given the cost of college, this could be quite a benefit, indeed.

- **Satisfy Requirements.** Some colleges will allow you to "place out" of certain requirements if you do well on an AP Exam, even if they do not give you actual college credits. For example, you might not need to take an introductory-level course, or perhaps you might not need to take a class in a certain discipline at all.

- **Admissions Plus.** Even if your AP Exam will not result in college credit or even allow you to place out of certain courses, most colleges will respect your decision to push yourself by taking an AP Course or even an AP Exam outside of a course. A high score on an AP Exam shows mastery of more difficult content than is taught in many high school courses, and colleges may take that into account during the admissions process.

OTHER RESOURCES

There are many resources available to help you improve your score on the AP Biology Exam, not the least of which are your teachers. If you are taking an AP class, you may be able to get extra attention from your teacher, such as obtaining feedback on your essays. If you are not in an AP course, reach out to a teacher who teaches AP Biology, and ask if the teacher will review your essays or otherwise help you with content.

Another wonderful resource is **AP Central**, the official site of the AP Exams. The scope of the information at this site is quite broad and includes

- Course Description, which includes details on what content is covered and sample questions
- Sample questions from the AP Biology exam
- Free-response question prompts and multiple-choice questions from previous years

The AP Central home page address is: **http://apcentral.collegeboard.com/home.**

For up-to-date information about the ongoing changes to the AP Biology Exam Course, please visit: http://apcentral.collegeboard.com/apc/public/courses/teachers_corner/2117.html.

Finally, **The Princeton Review** offers tutoring and small group instruction. Our expert instructors can help you refine your strategic approach and add to your content knowledge. For more information, call 1-800-2REVIEW.

DESIGNING YOUR STUDY PLAN

As part of the Introduction, you identified some areas of potential improvement. Let's now delve further into your performance on Test 1, with the goal of developing a study plan appropriate to your needs and time commitment.

Read the answers and explanations associated with the Multiple-Choice questions (starting at page 31). After you have done so, think about the following items:

- Review the Overview of Content Topics on pages 44 and 45 and, next to each one, indicate your rank of the topic as follows: "1" means "I need a lot of work on this," "2" means "I need to beef up my knowledge," and "3" means "I know this topic well."

- How many days/weeks/months away is your AP Biology Exam?

- What time of day is your best, most focused study time?

- How much time per day/week/month will you devote to preparing for your AP Biology Exam?

- When will you do this preparation? (Be as specific as possible: Mondays and Wednesdays from 3 to 4 P.M., for example)

- What are your overall goals in using this book?

Part III
Test-Taking Strategies for the AP Biology Exam

1   How to Approach Multiple-Choice Questions
2   How to Approach Free-Response Questions

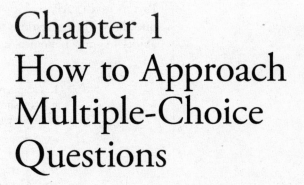

Chapter 1
How to Approach
Multiple-Choice
Questions

SECTION I

As we mentioned earlier, the multiple-choice section consists of the following two parts:

- Part A—contains regular multiple-choice questions and multiple-choice questions dealing with an experiment or a set of data
- Part B—contains grid-in questions

Part A

Part A of the AP Biology Exam consists of 63 run-of-the-mill multiple-choice questions. These questions test your grasp of the fundamentals of biology and your ability to apply biological concepts to help problem-solve. Here's an example.

22. If a segment of DNA reads 5´-ATG-CCA-GCT-3´, the mRNA strand that results from the transcription of this segment will be

(A) 3´-TAC-GGT-CGA-5´
(B) 3´-UAC-AGT-CAA-5´
(C) 3´-TAA-GGU-CGA-5´
(D) 3´-TAC-GGT-CTA-5´

Don't worry about the answer to this question. The majority of the questions in Part A are presented in this format. A few questions may include a figure, a diagram, or a chart.

The second part of the first portion also consists of multiple-choice questions, yet here you're asked to think logically about different biological experiments or data. The next page shows a typical example.

Questions 60 and 61 refer to the following diagram and information.

To understand the workings of neurons, an experiment was conducted to study the neural pathway of a reflex arc in frogs. A diagram of a reflex arc is given below.

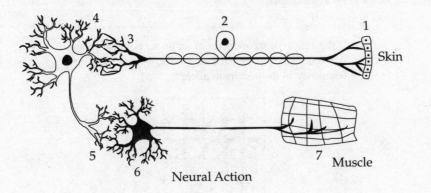

Neural Action

60. Which of the following represents the correct pathway taken by a nerve impulse as it travels from the spinal cord to effector cells?

(A) 1-2-3-4
(B) 6-5-4-3
(C) 2-3-4-5
(D) 4-5-6-7

61. The brain of the frog is destroyed. A piece of acid-soaked paper is applied to the frog's skin. Every time the piece of paper is placed on its skin, one leg moves upward. Which of the following conclusions is best supported by the experiment?

(A) Reflex actions are not automatic.
(B) Some reflex actions can be inhibited or facilitated.
(C) All behaviors in frogs are primarily reflex responses.
(D) This reflex action bypasses the central nervous system.

You'll notice that these particular questions refer to an experiment. Many of the questions in this portion test your ability to integrate information, interpret data, and draw conclusions from the results.

Part B

Part B consists of six grid-in questions where an answer needs to be calculated based on information presented in the question. That numeric response is then filled in on a grid and bubbled accordingly. Answers can be in the form of integers, decimals, or fractions. A four-function calculator can be used on these questions. Here's a typical example.

34. If the genotype frequencies of an insect population are AA = 0.49, Aa = 0.42, and aa = 0.09, what is the gene frequency of the dominant allele?

We mentioned earlier that our approach is strategy-based. As you're about to see, many of these strategies are based on common sense—for example, using mnemonics like "ROY G. BIV." (Remember that one? It's the mnemonic for red, orange, yellow, green, blue, indigo, violet—the colors of the spectrum.) Others are not so common-sensical. In fact, we're going to ask you to throw out much of what you've been taught when it comes to taking standardized tests.

Pace Yourself

When you take a test in school, how many questions do you answer? Naturally, you try to answer all of them. You do this for two reasons: (1) Your teacher told you to, and (2) if you left a question blank, your teacher would mark it wrong. However, that's not the case when it comes to the AP Biology Exam. In fact, finishing the test is the worst thing you can do. Before we explain why, let's talk about timing.

One of the main reasons that taking the AP Biology Exam is so stressful is the time constraint we discussed above—75 seconds per multiple-choice question and 21 minutes per essay. If you had all day, you would probably do much better. We can't give you all day, but we can do the next best thing: We can give you more time for each question. How? By having you slow down and answer fewer questions.

Slowing down, and doing well on the questions you do answer, is the best way to improve your score on the AP Biology Exam. Rushing through questions in order to finish, on the other hand, will always hurt your score. When you rush, you're far more likely to make careless errors, misread, and fall into traps. Keep in mind that blank answers are not counted against you.

The Three-Pass System

The AP Biology Exam covers a broad range of topics. There's no way, even with our extensive review, that you will know everything about every topic in biology. So what should you do?

Do the Easiest Questions First

The best way to rack up points is to focus on the easiest questions first. Many of the questions asked on the test will be straightforward and require little effort. If you know the answer, nail it and move on. Others, however, will not be presented in such a clear way. As you read each question, decide if it's easy, medium, or hard. During a first pass, do all the easy questions. If you come across a problem that seems time-consuming or completely incomprehensible, skip it. Remember,

> Easier questions count just as much as harder ones, so your time is better spent on shorter, easier questions.

Save the medium questions for the second pass. These questions are either time-consuming or require that you analyze all the answer choices. If you come across a question that makes no sense from the outset, save it for the last pass. You're far more likely to fall into a trap, or settle on a silly answer.

Watch Out for Those Bubbles!

Because you're skipping problems, you need to keep careful track of the bubbles on your answer sheet. One way to accomplish this is by answering all the questions on a page and then transferring your choices to the answer sheet. If you prefer to enter them one by one, make sure you double-check the number beside the ovals before filling them in. We'd hate to see you lose points because you forgot to skip a bubble!

So then, what about the questions you don't skip?

Process of Elimination (POE)

On most tests, you need to know your material backward and forward to get the right answer. In other words, if you don't know the answer beforehand, you probably won't answer the question correctly. This is particularly true of fill-in-the-blank and essay questions. We're taught to think that the only way to get a question right is by knowing the answer. However, that's not the case on Section I of the AP Biology Exam. You can get a perfect score on this portion of the test without knowing a single right answer, provided you know all the wrong answers!

What are we talking about? This is perhaps the single most important technique in terms of the multiple-choice section of the exam. Let's take a look at the following example:

1. The structures that act as the sites of gas exchange in a woody stem are the

   (A) lungs
   (B) lenticels
   (C) ganglia
   (D) lentil beans

Now if this were a fill-in-the-blank-style question, you might be in a heap of trouble. But let's take a look at what we've got. You see "woody stem" in the question, which leads you to conclude that we're talking about plants. Right away, you know the answer is not (A) or (C) because plants don't have lungs or ganglia. Now we've got it down to (B) and (D). Notice that (B) and (D) are very similar. Obviously, one of them is a trap. At this point, if you don't know what "lentil beans" are, you have to guess. However, even if we don't know precisely what they are, it's safe to say that most of us know that lentil beans have nothing to do with plant respiration. Therefore, the correct answer is (B), lenticels.

Although our example is a little goofy and doesn't look exactly like the questions you'll be seeing on the test, it illustrates an important point.

> **Process of Elimination** is the best way to approach the multiple-choice questions.

Even if you don't know the answer right off the bat, you'll surely know that two or three of the answer choices are not correct. What then?

Aggressive Guessing

ETS tells you that random guessing will not affect your score. This is true. There is no guessing penalty on the AP Biology exam. For each correct answer you'll receive one point, and you will not lose any points for each incorrect answer.

Although you won't lose any points for wrong answers, you should guess aggressively by getting rid of the incorrect answer choices. The moment you've eliminated a couple of answer choices, your odds of getting the question right, even if you guess, are far greater. If you can eliminate as many as two answer choices, your odds improve enough that it's in your best interest to guess.

Word Associations

Another way to rack up the points on the AP Biology Exam is by using word associations in tandem with your POE skills. As you learn these key terms in your AP Biology course, make sure you group them by association because ETS is bound to ask about them on the AP Biology Exam. What do we mean by "word associations"? Let's take the example of mitosis and meiosis.

You'll soon see from our review that there are several terms associated with mitosis and meiosis. *Synapsis*, *crossing-over*, and *tetrads*, for example, are words associated with meiosis but not mitosis. Take a look at the following question:

> 2. Which of the following typifies cytokinesis during mitosis?
>
> (A) Crossing-over
> (B) Formation of tetrads
> (C) Synapsis
> (D) Division of the cytoplasm

This might seem like a difficult problem. But let's think about the associations we just discussed. The question asks us about mitosis. However, answer choices (A), (B), and (C) all mention events that we've associated with meiosis. Therefore, they are out. Without even racking your brain, you've managed to find the correct answer choice: (D). Not bad!

Once again, don't worry about the science for now. What is important to recognize is that by combining the associations we'll offer throughout this book and your aggressive POE techniques, you'll be able to rack up points on problems that might have seemed difficult at first.

Mnemonics—or the Biology Name Game

One of the big keys to simplifying biology is the organization of terms into a handful of easily remembered packages. The best way to accomplish this is by using mnemonics. Biology is all about names: the names of chemical structures, processes, theories, and so on. How are you going to keep them all straight? A mnemonic, as you may already know, is a convenient device for remembering something.

For example, one important issue in biology is taxonomy, that is, the classification of life forms, or organisms. Organisms are classified in a descending system of similarity, leading from kingdoms (the broadest level) to species (the most specific level). The complete order runs: domain, kingdom, phylum, class, order, family, genus, and species. Don't freak out yet. Look how easy it becomes with a mnemonic.

King Philip of Germany decided to walk to America. What do you think happened?

Dumb	→	Domain
King	→	Kingdom
Philip	→	Phylum
Came	→	Class
Over	→	Order
From	→	Family
Germany	→	Genus
Soaked	→	Species

Learn the mnemonic and you'll never forget the science!

Mnemonics can be as goofy as you like, so long as they help you remember. Remember: The important thing is that you remember the information, not how you remember it.

Identifying EXCEPT Questions

About 10 percent of the multiple-choice questions in Section I are EXCEPT/NOT/LEAST questions. With this type of question, you must remember that you're looking for the *wrong* (or the least correct) answer. The best way to go about these is by using POE.

More often than not, the correct answer is a true statement but is wrong in the context of the question. However, the other three tend to be pretty straightforward. Cross off the four that apply and you're left with the one that does not. Here's a sample question.

17. All of the following are true statements about gametes
    EXCEPT

    (A) they are haploid cells.
    (B) they are produced only in the reproductive
        structures.
    (C) they bring about genetic variation among
        offspring.
    (D) they develop from polar bodies.

If you don't remember anything about gametes and gametogenesis, or the production of gametes, this might be a particularly difficult problem. However, you may recall from your course reveiw that gametes are the "sex cells" of sexually reproducing organisms. As such, we know that they are haploid and are produced in the sexual organs. We also know that they come together to create offspring.

From this very basic review, we know immediately that (A) and (B) are not our answers. Both of these are accurate statements, so we eliminate them. That leaves us with (C) and (D). If you have no idea what (D) means, focus on (C). In sexual reproduction, each parent contributes one gamete, or half the genetic complement of the offspring. This definitely helps vary the genetic makeup of the offspring. Answer choice (C) is a true statement, so it can be eliminated. The correct answer is (D).

Remember: The best way to answer these types of questions is to spot all the right statements and cross them off. You'll wind up with the wrong statement, which happens to be the correct answer

REFLECT

Respond to the following questions:

- How long will you spend on multiple-choice questions?

- How will you change your approach to multiple-choice questions?

- What is your multiple-choice guessing strategy?

- Will you seek further help outside of this book (such as a teacher, tutor, or AP Central) on how to approach the questions that you will see on the AP Biology exam?

Chapter 2
How to
Approach Free-
Response Questions

THE ART OF THE ETS ESSAY

You are given two essay questions and six short-form free-response questions to answer in 80 minutes. The best way to rack up points on this section is to give the essay readers what they're looking for. Fortunately, we know precisely what that is.

The ETS essay reviewers have a checklist of key terms and concepts that they use to assign points. We like to call these "hot button" terms. Quite simply put, for each hot button that you include in your essay, you will receive a predetermined number of points. For example, if the essay question deals with the function of enzymes, the ETS graders are instructed to give 2 points for a mention of the "lock-and-key theory of enzyme specificity."

Naturally, you can't just compose a "laundry list" of scientific terms. Otherwise, it wouldn't be an essay. What you can do, however, is organize your essay around a handful of these key, or hot button, points. The most effective and efficient way to do this is by using the 10-minute reading period to brainstorm and come up with the scientific terms. Then outline your essay before you begin to write, using your hot buttons as your guide.

READ THE QUESTIONS CAREFULLY

ETS gives you 10 minutes to read the questions and organize your thoughts before you begin writing. If you use these 10 minutes wisely, you can breeze your way through the essays. The first thing you should do is take less than a minute to skim all of the questions, and put them into your own personal order of difficulty from easiest to toughest. Once you've decided the order in which you will answer the questions (easiest first, hardest last), you can begin to formulate your responses. Your first step should be a more detailed assessment of each question.

The most important advice we can give you is to read each question at least twice. As you read the question, focus on key words, especially "direction words." Almost every essay question begins with a direction word. Some examples of direction words are *discuss*, *define*, *explain*, *describe*, *compare*, and *contrast*. If a question asks you to discuss a particular topic, you should give a viewpoint, and support it with examples. If a question asks you to compare two things, you should discuss how the two things are similar. On the other hand, if the question asks you to contrast two things, you need to show how these things are different.

Many students lose points on their essays because they either misread the question or fail to do what's asked of them.

BRAINSTORM

Your next objective during the 10-minute preview should be to organize your thoughts.

Once you've read the questions, you need to brainstorm. Jot down as many key terms and concepts as you can. Don't forget, the test reviewers assign points on the basis of these key concepts: For each one that you mention and/or explain, you get a point.

How many do you need? You won't need all of them, that's for sure. However, you will need enough to get you the maximum number of points for that question. Most of these you can pull directly from your reading in this book. Remember the "Key Words" lists in each chapter of this book? They are essential to your preparation for this part of the test. Use your word associations to generate thorough lists. Once you've come up with as many different hot buttons as possible, you're ready to leap into your outline.

Don't spend more than about 2 minutes per question brainstorming. Once your 10-minute reading period is up you should be ready to start writing your essays.

OUTLINE YOUR ESSAY

Have you ever written yourself into a corner? You're halfway through your essay when you suddenly realize that you have no idea whatsoever where you're going with your train of thought. To avoid this (and the panic that accompanies it), take a few minutes to draft an outline.

Your outline should incorporate as many of the hot buttons as you need in order to maximize your score. In other words, if they ask for two examples, choose only those two with which you are most comfortable. In your outline, make notes about the crucial points to mention with regard to each topic or key word. Once your outline is complete, you're ready to move on to writing the actual essay.

All of this preparation may seem time-consuming. However, it should take no more than four or five minutes per essay. What's more, it will greatly simplify the whole essay-writing process. So while you lose a little time at the outset, you'll more than make up for it when it comes time to actually write your answer.

DEVELOP YOUR IDEAS IN EACH ESSAY

Now you can use your outline to write your essay. Most students can come up with key terms or phrases that concern a particular biology topic. What separates a high-scoring student from a low-scoring student is *how* the student develops his or her thoughts on each essay. Besides giving the hot buttons, you'll need to elaborate on your thoughts and ideas. For example, don't just throw out a list of terms that pertain to meiosis and mitosis (like *synapsis*, *crossing-over*, and *gametes*). Go one step further. Make sure you mention the *significance* of meiosis (i.e., it produces genetic variability). This extra piece of information will earn you an extra point.

Generally, you'll need to write about two to four paragraphs, depending on the number of parts contained in each essay question. In addition, be sure to give the appropriate number of examples for each essay. If the question asks for three examples, give only three examples. If you present more than is required, the test reviewers won't even read them or count them toward your score. The bottom line is this: Stick to the question.

Answer Each Part of the Essay Question Separately

The more parts there are to an essay question, the more important it is to pace yourself. On each essay, you're better off writing a little bit for each part than you are spending all your time on any one part of a question. Why? Even if you were to write the perfect answer to one part of a question, there's a limit to the number of points the test reviewers can assign to that part. Moreover, by writing a separate paragraph for each section, you make the test grader's job that much easier. When test readers have an easy time reading your essay, they're more likely to award points: It comes across as clearer and more organized. Readers have also requested that students label each part of their essay answers, so write "Part a", "Part b", and so on, accordingly in your response.

Finally, don't spend too much time writing a fancy introduction. You won't get brownie points for beautifully written openings like, "It was the best of experiments, it was the worst of experiments." Just leap right into the essay. And don't worry too much about grammar or spelling errors. Your grammar can hurt only if it's so bad that it seriously impairs your ability to communicate.

Incorporate Elements of an Experimental Design

Because one of the essay questions will be experimentally based, you'll need to know how to design an experiment. Most of these questions require that you present an appropriately labeled diagram or graph. Otherwise, you'll only get partial credit for your work.

There are two things you must remember when designing experiments on the AP Biology Exam: (1) Always label your figures, and (2) include controls in all experiments. Let's take a closer look at these two points.

Know How to Label Diagrams and Figures

Let's briefly discuss the important elements in setting up a graph. The favorite type of graph on the AP Biology Exam is the *coordinate graph*. The coordinate graph has a horizontal axis (*x*-axis) and a vertical axis (*y*-axis).

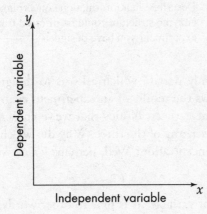

The *x*-axis usually contains the *independent variable*—the thing that's being manipulated or changed. The *y*-axis contains the *dependent variable*—the thing that's affected when the independent variable is changed.

Now let's look at what happens when you put some points on the graph. Every point on the graph represents both an independent variable and a dependent variable.

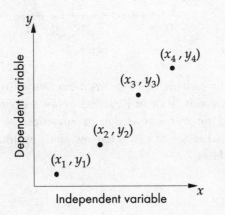

Once you draw both axes and label the axes as *x* and *y*, you can plot the points on the graph. Let's look at the following question.

1. Enzymes are biological catalysts.

    a. Relate the chemical structure of an enzyme to its catalytic activity and specificity.
    b. Design an experiment that investigates the influence of temperature, substrate concentration, or pH on the activity of an enzyme.
    c. Describe what information concerning enzyme structure could be inferred from the experiment you have designed.

For now, let's discuss only part b, which asks us to design an experiment. Let's set up a graph that shows the results of an experiment examining the relationship between pH and enzyme activity. Notice that we've chosen only one factor here, pH. We could have chosen any of the three. Why did we choose pH and not temperature or substrate concentration? Well, perhaps it's the one we know the most about.

What is the independent variable? It is pH. In other words, pH is being manipulated in the experiment. We'll therefore label the *x*-axis with pH values from 0 through 14.

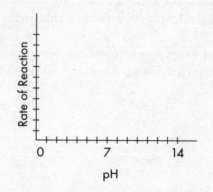

What is the dependent variable? It's the enzyme activity—the thing that's affected by pH. Let's label the *y*-axis "Rate of Reaction." Now we're ready to plot the values on the graph. Based on our knowledge of enzymes, we know that for most enzymes the functional range of pH is narrow, with optimal performance occurring at or around a pH of 7.

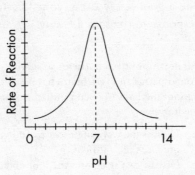

Now you should interpret your graph. If the pH level decreases from a neutral pH of 7, the reaction rate of the enzyme will decrease. If the pH level increases, the rate of reaction will also decrease. Don't forget to include a simple explanation of your graph.

Include Controls in Your Experiments

Almost every experiment will have at least one variable that remains constant throughout the study. This is called the *control*. A control is simply a standard of comparison. What does a control do? It enables the biologist to verify that the outcome of the study is due to changes in the independent variable and nothing else.

Let's say the principal of your school thinks that students who eat breakfast do better on the AP Biology Exam than those who don't eat breakfast. He gives a group of 10 students from your class free breakfast every day for a year. When the school year is over, he administers the AP Biology Exam, and they all score brilliantly! Did they do well because they ate breakfast every day? We don't know for sure. Maybe the principal handpicked the smartest kids in the class to participate in the study.

In this case, the best way to be sure that eating breakfast made a difference is to have a control group. In other words, he would need to pick students in the class who *never* eat breakfast and follow them for a year. At the end of that year, he could send them in to take the AP Biology Exam. If they do just as well as the group that ate breakfast, then we can probably conclude that eating breakfast wasn't the only factor leading to higher AP scores. The group of students that didn't eat breakfast is called the control group because those students were not "exposed" to the variable of interest—in this case, breakfast.

Let's see if you can write a good essay using your free-response techniques. Take 21 minutes to write a response to the sample essay.

1. All organisms need nutrients to survive. Angiosperms and vertebrates have each developed various methods to obtain nutrients from their environment.

    a. Discuss the ways angiosperms and vertebrates procure their nutrients.
    b. Discuss two structures used for obtaining nutrients among angiosperms. Relate structure to function.
    c. Discuss two examples of symbiotic relationships that have evolved between organisms to obtain nutrients.

REFLECT

Respond to the following questions:

- How much time will you spend on the free-response questions?

- What will you do before you begin writing your free-response answers?

- Will you seek further help outside of this book (such as a teacher, tutor, or AP Central) on how to approach the questions that you will see on the AP Biology exam?

Part IV
Drills

3   Big Idea 1 Drill 1
4   Big Idea 1 Drill 1 Answers and Explanations
5   Big Idea 2 Drill 1
6   Big Idea 2 Drill 1 Answers and Explanations
7   Big Idea 3 Drill 1
8   Big Idea 3 Drill 1 Answers and Explanations
9   Big Idea 4 Drill 1
10 Big Idea 4 Drill 1 Answers and Explanations
11 Big Idea 1 Drill 2
12 Big Idea 1 Drill 2 Answers and Explanations
13 Big Idea 2 Drill 2
14 Big Idea 2 Drill 2 Answers and Explanations
15 Big Idea 3 Drill 2
16 Big Idea 3 Drill 2 Answers and Explanations
17 Big Idea 4 Drill 2
18 Big Idea 4 Drill 2 Answers and Explanations
19 Big Idea 1 Drill 3
20 Big Idea 1 Drill 3 Answers and Explanations
21 Big Idea 2 Drill 3
22 Big Idea 2 Drill 3 Answers and Explanations
23 Big Idea 3 Drill 3
24 Big Idea 3 Drill 3 Answers and Explanations
25 Big Idea 4 Drill 3
26 Big Idea 4 Drill 3 Answers and Explanations
27 Big Idea 1 Drill 4
28 Big Idea 1 Drill 4 Answers and Explanations
29 Big Idea 2 Drill 4
30 Big Idea 2 Drill 4 Answers and Explanations
31 Big Idea 3 Drill 4
32 Big Idea 3 Drill 4 Answers and Explanations
33 Big Idea 4 Drill 4
34 Big Idea 4 Drill 4 Answers and Explanations
35 Free-Response Short Answer Drill
36 Free-Response Short Answer Drill Answers and Explanations
37 Free-Response Long Answer Drill
38 Free-Response Long Answer Drill Answers and Explanations

Chapter 3
Big Idea 1 Drill 1

BIG IDEA 1 DRILL 1

Multiple-Choice Questions

1. The property of water that allows many organisms with similar body compositions of water to maintain a constant body temperature is
   (A) polarity
   (B) adhesion
   (C) cohesion
   (D) heat capacity

2. All living organisms are made up of chemical compounds containing a skeleton of atoms with which of the following properties?
   I. Atoms that cannot bind with carbon
   II. Atoms that can bind with nitrogen
   III. Atoms that can bind with oxygen
   (A) I only
   (B) I and II only
   (C) II and III only
   (D) I, II, and III

3. The first living organisms are thought to have primarily derived from all of the following EXCEPT
   (A) ammonia
   (B) hydrogen
   (C) oxygen
   (D) water

4. Some scientists theorize that mitochondria were derived from simple prokaryotic cells. Which of the following could explain this theory?
   (A) A eukaryotic cell underwent exocytosis of a prokaryotic cell
   (B) A eukaryotic cell underwent endocytosis of ATP
   (C) A prokaryotic cell underwent exocytosis of ATP
   (D) A eukaryotic cell underwent endocytosis of a prokaryotic cell

5. Both prokaryotic and eukaryotic cells contain all of the following EXCEPT

(A) organelles
(B) DNA
(C) plasma membrane
(D) ribosomes

6. Which of the following elements is essential to most cellular processes, including active transport, protein synthesis, and glycolysis?

(A) Oxygen
(B) Phosphorus
(C) Nitrogen
(D) Iron

7. In the figure, the enzyme to the right of the arrow is inactivated due to

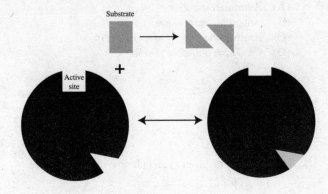

(A) allosteric activation
(B) allosteric inhibition
(C) competitive inhibition
(D) denaturation

8. The two molecules that result from glycolysis are each made up of how many carbons?

(A) Two
(B) Three
(C) Six
(D) Twelve

9. In the absence of oxygen, both prokaryotes and eukaryotes can utilize which process to produce ATP?

(A) Aerobic respiration
(B) Anaerobic respiration
(C) Krebs cycle
(D) Oxidative phosphorylation

10. Both cellular respiration and photosynthesis produce

  I. ATP
  II. Coenzyme A
  III. Electron carriers

(A) I only
(B) I and III only
(C) II and III only
(D) I, II, and III

11. The products of photosynthesis are similar to the

(A) products of cellular respiration
(B) reactants of cellular respiration
(C) products of the Krebs cycle
(D) reactants of the Krebs cycle

12. The hereditary information for all living organisms consists of a backbone of nucleotides linked by

(A) hydrogen bonds
(B) purine and pyrimidine bases
(C) nitrogenous bonds
(D) phosphodiester bonds

13. What is the complementary mRNA strand that would be produced using this piece of DNA as a template: TTCATGCAA?

    (A) AAGTACGTT
    (B) AAGUACGUU
    (C) UUCAUGCAA
    (D) UUGAUGCAA

14. Which type of mutation would result in the biggest deficit to an organism?

    (A) Single base pair deletion
    (B) Three base pair deletion
    (C) Missense mutation
    (D) Silent mutation

15. In mitosis, the total number of chromatids doubles during

    (A) G1 stage
    (B) S phase
    (C) G2 stage
    (D) prophase

16. Which process(es) help(s) to drive genetic diversity?

    I.   Mutation
    II. Mitosis
    III. Meiosis

    (A) I only
    (B) II only
    (C) I and II only
    (D) I and III only

17. Which of the following is NOT true of haploid cells?

    (A) Haploid cells have half the number of chromosomes as diploid cells.
    (B) In humans, haploid cells have 23 chromosomes.
    (C) One example of haploid cells are human gametes.
    (D) The two cells resulting from mitosis are haploid cells.

18. In mice, the long tail allele (T) is dominant over the short tail allele (t). A long tailed mouse is crossed with a short tailed mouse. Half of their offspring have long tails and half have short tails. What are the genotypes of the two crossed mice?

    (A) TT and tt
    (B) Tt and Tt
    (C) Tt and tt
    (D) tt and tt

19. Which of the following is NOT an example of Mendelian inheritance?

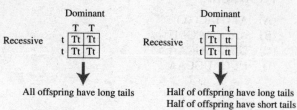

All offspring have long tails

Half of offspring have long tails
Half of offspring have short tails

    (A) Heterozygote individuals display the dominant phenotype.
    (B) Homozygote recessive individuals display the recessive phenotype.
    (C) Heterozygote individuals display an intermediate phenotype.
    (D) Different traits separate and recombine independently of other traits.

20. Which type of evolutionary selection is displayed by the graph below?

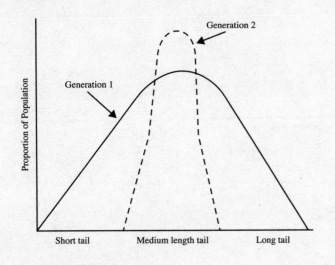

(A) Circular selection
(B) Directional selection
(C) Disruptive selection
(D) Stabilizing selection

21. A dog that learns to stay within the confines of a yard to avoid being shocked by an electric fence is an example of

(A) imprinting
(B) insight
(C) classical conditioning
(D) operant conditioning

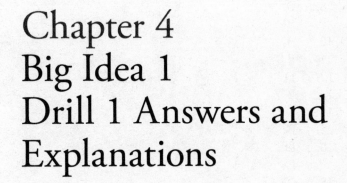

Chapter 4
Big Idea 1
Drill 1 Answers and
Explanations

ANSWER KEY

1. D
2. C
3. C
4. D
5. A
6. B
7. B
8. B
9. B
10. B
11. B
12. D
13. B
14. A
15. B
16. D
17. D
18. C
19. C
20. D
21. D

ANSWERS AND EXPLANATIONS

Multiple-Choice Questions

1. **D** Organisms that are made up mostly of water are able to keep a near constant body temperature because water has a very high heat capacity. This means that it requires a great amount of heat energy to increase the temperature of water. Water is also polar, but this means that it has partially positive and partially negative charges, and has no effect on body temperature (A). Adhesion and cohesion are also properties of water, but the ability of water to stick to other water molecules and other substances would not affect body temperature eliminating (B) and (C).

2. **C** All living organisms are made up of chemical compounds with a carbon chain backbone. Carbon molecules CAN form a bond with other carbon molecules, therefore I is false; eliminate (A), (B), and (D). Carbon molecules can form a bond with nitrogen as well as oxygen.

3. **C** Ammonia (A), hydrogen (B), and water (D) are all thought to be components of the earth's primitive atmosphere and were used in experiments to derive organic components; oxygen (C) is not thought to be a major component of this early atmosphere. Furthermore, early autotrophs are thought to be responsible for the oxygen of earth's atmosphere, so there would not be significant oxygen prior to the existence of living organisms.

4. **D** Mitochondria are organelles found in eukaryotic organisms, so if mitochondria were derived from prokaryotic cells, a prokaryotic cell would have to have entered a eukaryotic cell. Exocytosis is the excretion of something out of a cell, so eliminate (A) and (C). While ATP is produced by mitochondria, an organelle must be membrane-bound and the stem tells us that mitochondria is derived from prokaryotes, not simple ATP (B).

5. **A** Prokaryotic and eukaryotic cells are both surrounded by a plasma membrane (C), and both contain DNA as their genetic material (B) and ribosomes for protein synthesis (D). Only eukaryotic cells contain organelles.

6. **B** Adenosine triphosphate (ATP) is the energy source for the cell and is required for many activities of the cell, including active transport, protein synthesis, and glycolysis. Three phosphate molecules carry the energy of ATP. While oxygen is required for some cellular processes, including the Krebs cycle and oxidative phosphorylation, it is not an energy source and is not required for the processes listed (A). Nitrogen is one of the essential elements of life but is not an energy source for cellular processes (C). Iron is a trace element found in the body in very small quantities (D).

7.   **B**   The figure depicts an example of feedback inhibition, where the product of the reaction catalyzed by the enzyme inhibits the enzyme by binding at an allosteric site (a site other than the active site). Allosteric activation would occur if the product activated, or turned on, the enzyme (A). Competitive inhibitors compete with substrate for binding at the active site (C). Denaturation occurs when an enzyme is inactivated by heat (D).

8.   **B**   The end product of glycolysis is two molecules of pyruvate, a three-carbon molecule. Glucose, the reactant for glycolysis, is a six-carbon molecule (C), and it is split in half during the course of the reactions in glycolysis, eliminating (A) and (D).

9.   **B**   Anaerobic respiration is the process by which cells can produce energy when oxygen is not available, and both prokaryotes and eukaryotes have the ability to carry out aerobic respiration. Aerobic respiration, by definition, requires oxygen (A). While the Krebs cycle does not directly utilize oxygen, it will not occur in the absence of oxygen (C). Oxidative phosphorylation utilizes oxygen as its final electron acceptor, and thus cannot occur in the absence of oxygen (D).

10. **B**   The overall goal of both cellular respiration and photosynthesis is to produce energy as ATP; therefore, eliminate (C) as it does not include I. Coenzyme A is required to produce Acetyl CoA from pyruvate in cellular respiration; it does not play a role in photosynthesis, so eliminate (D). Electron carriers are produced in both processes, specifically NADH and $FADH_2$ in cellular respiration, and NADPH in photosynthesis; eliminate (A).

11. **B**   The products of photosynthesis are glucose ($C_6H_{12}O_6$) and oxygen, which are the reactants of cellular respiration. The products of cellular respiration are carbon dioxide, water, and ATP (B). Acetyl CoA enters the Krebs cycle and are ultimately converted to CO_2 as ATP, and electron carriers are produced along the way (C) and (D).

12. **D**   DNA contains the hereditary information for all living organisms and is made up of individual nucleotides linked via two phosphoester bonds, called a phosphodiester bond. Hydrogen bonds link two strands of DNA via purine-pyrimidine base pairs to make its double-stranded structure, so eliminate (A) and (B). The bases are nitrogen based, but these nitrogens do not connect nucleotides together (C).

13. **B**   One of the main differences between DNA and RNA is that DNA contains thymine as a base, while RNA instead contains uracil; eliminate (A). The complimentary strand of mRNA must contain complementary bases that would base pair with the DNA strand—therefore, C would replace G, G would replace C, A would replace T, and U would replace A, consistent with choice (B).

14. **A**   A single base pair deletion would result in a different sequence of codons following the deletion than originally present, thus causing a "frameshift" mutation—a shift in the entire reading frame for the DNA. Because each subsequent codon would be different, the resulting protein from translation would be completely different amino acids following the site of the mutation. It is unlikely that a protein could maintain its function given a different sequence of amino acids, so a single base pair deletion is likely to be very detrimental to an organism. A three base pair deletion would

remove one complete codon from the DNA sequence but would not shift the reading frame—thus, only one amino acid would be lost, but the remaining amino acid sequence would be correct, and protein function is likely to be at least partially spared; eliminate choice (B). A missense mutation means that one amino acid is replaced by a new amino acid—again, while one amino acid will be different in the resulting protein, it is unlikely to have a completely devastating effect on protein function, so eliminate choice (C).

15. **B** S phase stands for "Synthesis," meaning DNA synthesis. This is the stage of interphase in which every chromosome in the nucleus of a cell is duplicated, resulting in sister chromatids held together by a centromere. G1 and G2 stand for "Gap", and are the periods in which the cell grows and prepares for mitosis, (A) and (C). Interphase is the first stage of mitosis in which chromosomes condense and the mitotic spindle assembles (D).

16. **D** Both mutation and meiosis help to create genetic diversity within a population. Random mutation results in brand new type of a gene that causes a new trait in a population. If this new trait does not cause the organism to die, it can pass on this new trait to its offspring, and thus the population increases in diversity. I does drive genetic diversity, so eliminate (B). Mitosis simply produces daughter cells which are exactly identical to the parent cell—because it is just an increase in number with no changes to genes or arrangements of genes, II does not drive genetic diversity, so eliminate (C). Meiosis is responsible for gamete formation, and involves crossing-over, a process by which different versions of genes are swapped between homologous chromosomes, for a new arrangement of different versions of genes. Therefore, meiosis does lead to genetic diversity—eliminate choice (A).

17. **D** Haploid cells have only one set of chromosomes, as opposed to diploid cells, which have two sets of chromosomes. Thus, haploid cells have half as many chromosomes as diploid cells, so you can eliminate choice (A) because it is true. A typical diploid human cell has 46 chromosomes, so a haploid human cell has 23 chromosomes (B). Human gametes are haploid cells, with 23 chromosomes each (C). Mitosis produces daughter cells exactly identical to the diploid parent cell, and thus the products of mitosis are both diploid cells, so choice (D) is false.

18. **C** This problem is an example of a test cross, where a dominant phenotype individual is crossed with a recessive phenotype individual. You know that the recessive phenotype is homozygous recessive (tt)—that is the only way to have a short tail. Eliminate choice (B) because it does not include tt. Based on its appearance, you can't know if the long-tailed mouse is TT or Tt, but you know it must have at least one dominant allele; it cannot be tt or else it would have a short tail. So eliminate choice (D). Because some of the offspring have short tails and some have long tails, both parents must have one of the recessive genes (t). Eliminate choice (A).

19. **C** Mendelian inheritance is the basis for classical dominance in which one copy of a dominant allele produces the dominant phenotype (A). Homozygous recessive individuals are the only ones who will display the recessive phenotype (B). Heterozygotes have one copy of the dominant allele and one copy of the recessive allele, and thus would display the dominant phenotype in Mendelian

inheritance, so choice (C) is NOT and example of Mendelian inheritance. One of Mendel's laws is the law of independent assortment which states the traits can segregate and recombine independently of other traits (D).

20. **D** In the graph, Generation 2 moves more toward the most common trait of medium length tails, with less of the population having shorter or longer tails. This is an example of stabilizing selection, in which extreme traits are eliminated in subsequent generations. There is no such thing as circular selection (A). Directional selection favors one phenotype at one of the extremes of the distribution (B). Disruptive selection occurs when both extremes (short and long tails) are favored, and the common trait (medium length tail) is less represented in subsequent generations (C).

21. **D** Operant conditioning is also known as trial-and-error learning, and it occurs when an animal learns to perform an act in order to receive a reward or to avoid a behavior in order to avoid a punishment. Imprinting is how an animal learns to recognize members of its own species and usually occurs early in its life (A). Insight is the ability to use higher-order thinking to solve problems and is only demonstrated by humans and other higher-order primates (B). Classical conditioning was classically described by Pavlov ringing a bell each time he fed his dogs food, which led to them eventually salivating whenever he rang the bell (regardless of if he gave them food). Repeated instances of an event linked to a stimulus led to a predictable outcome in the animals when the stimulus was applied (C).

Chapter 5
Big Idea 2 Drill 1

BIG IDEA 2 DRILL 1

Multiple-Choice Questions

1. Based on the chart below, a basic unit of malt sugar could be made similar to a basic unit of corn syrup through the process of

Sugar name	Carbohydrate	Basic unit
Brown sugar	Sucrose	Disaccharide
Corn syrup	Glucose	Monosaccharide
High-fructose corn syrup	Fructose	Monosaccharide
Malt sugar	Maltose	Disaccharide
Milk sugar	Lactose	Disaccharide

(A) condensation
(B) dehydration synthesis
(C) hydrolysis
(D) hydrogen bonding

2. Which of the following functional groups combine to form peptide bond?

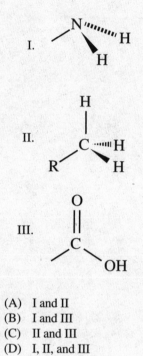

(A) I and II
(B) I and III
(C) II and III
(D) I, II, and III

3. According to the chart below, the lipid membrane-spanning portion of a transmembrane protein is most likely to be made up of

Amino Acids Reference Chart

Very hydrophobic	Hydrophobic	Neutral	Hydrophilic
Leucine Isoleucine Phenylalanine Tryptophan Valine Methionine	Cysteine Tyrosine Alanine	Threonine Glutamate Glycine Serine Glutamine Aspartate	Arginine Lysine Asparagine Histidine Proline

    I. Alanine
   II. Arginine
  III. Isoleucine

(A) I only
(B) II only
(C) I and III only
(D) II and III only

4. Which cell is likely to have the greatest number of mitochondria?

(A) Fat cell
(B) Hair cell
(C) Heart muscle cell
(D) Skin cell

5. Glycolysis is similar to oxidative phosphorylation in that both processes

(A) require ATP
(B) produce ATP
(C) require oxygen
(D) occur in the mitochondria

6. Why is the C_4 pathway for carbon fixation especially important for plants in climates such as the desert?

(A) PEP carboxylase has a high affinity for CO_2.
(B) PEP carboxylase has a low affinity for CO_2.
(C) RuBP carboxylase has a high affinity for CO_2.
(D) RuBP carboxylase has a low affinity for CO_2.

7. Which enzyme is necessary for completion of the lagging strand, but not the leading strand, in DNA synthesis?

   (A) Helicase
   (B) Polymerase
   (C) Ligase
   (D) RNA primase

8. The correct order in which enzymes are utilized to express a gene is

   (A) RNA polymerase – spliceosome – ribosome
   (B) ribosome – RNA polymerase – spliceosome
   (C) RNA polymerase – tRNA – ribosome
   (D) ribosome – RNA polymerase – tRNA

9. During S phase of mitosis, DNA polymerase is required for the duplication of DNA. When is DNA polymerase synthesized?

   (A) G1 stage
   (B) G2 stage
   (C) Prophase
   (D) DNA polymerase is not synthesized during the cell cycle.

10. Which is NOT a function of mitosis?

   (A) Asexual reproduction
   (B) Growth
   (C) Repair
   (D) Sexual reproduction

11. Which of the following is shown in the figure?

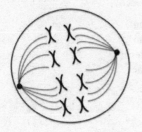

(A) Anaphase I of meisosis, in which sister chromatids separate at the centromere
(B) Metaphase I of meiosis, in which tetrads line up at the metaphase plate
(C) Metaphase of mitosis, in which replicated chromosomes line up randomly
(D G2 stage of interphase, immediately following the duplication of chromosomes

12. The organ primarily responsible for digestion of proteins, carbohydrates, and lipids is the

(A) esophagus
(B) stomach
(C) small intestine
(D) large intestine

13. The digestive enzyme secretin stimulates the pancreas to produce bicarbonate (HCO_3^-), which is released into the small intestine. What might be a reason for this series of events?

(A) Pancreatic bicarbonate helps to regulate the pH of the bloodstream.
(B) Pancreatic bicarbonate helps to neutralize stomach acid that enters the small intestine.
(C) Pancreatic bicarbonate is a waste product of protein digestion.
(D) Pancreatic bicarbonate increases the acidity of the small intestine to aid in digestion.

14. Through which heart valve(s) does oxygenated blood flow?

   I. Left atrioventricular (bicuspid) valve
  II. Right atrioventricular (tricuspid) valve
 III. Pulmonary semilunar valve

(A) I only
(B) I and II only
(C) I and III only
(D) II and III only

15. Which is a plausible explanation for the fact that peptide hormones have a faster onset of action than steroid hormones?

(A) Peptide hormones can diffuse across the plasma membrane of a target cell, while steroid hormones cannot.

(B) Steroid hormones bind to extracellular receptors, and these receptors take several hours to turn on.

(C) Peptide hormones bind to extracellular receptors and produce more immediate effects than steroid hormones, which bind DNA to affect transcription.

(D) Both steroid hormones and peptide hormones bind to DNA to affect transcription, but peptide hormones are able to get to the nucleus of the cell faster.

16. An egg fertilized by a sperm becomes a zygote, and goes through series of rapid cell divisions. Which of the following accurately portrays the correct order of embryonic development?

(A) Zygote → Morula →Gastrula → Blastula
(B) Zygote → Gastrula → Morula → Blastula
(C) Zygote → Gastrula → Blastula → Morula
(D) Zygote → Morula → Blastula → Gastrula

17. Which extraembryonic membrane provides energy for the growing embryo?

(A) Amnion
(B) Chorion
(C) Mesoderm
(D) Yolk sac

18. Which is NOT true regarding autotrophs and heterotrophs?

(A) Heterotrophs must get all of their energy from autotrophs.

(B) Autotrophs use basic building blocks, such as water and atmospheric gases, to make their own chemical energy.

(C) Heterotrophs break down the organisms they consume into organic substances.

(D) Autotrophs convert light energy to chemical energy via photosynthesis.

19. In glycolysis (simplified scheme shown below), the enzyme phosphofructokinase phosphorylates fructose-6-phosphate to produce fructose-1,6-bisphosphate. Which of the following might reasonably provide negative feedback inhibition on the phosphofructokinase enzyme?

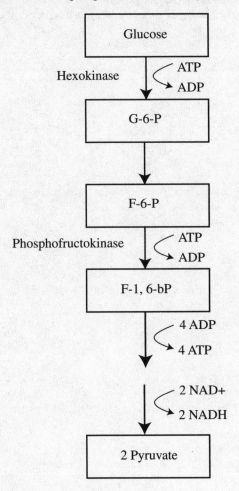

(A) ADP
(B) ATP
(C) Fructose-6-phosphate
(D) Glucose

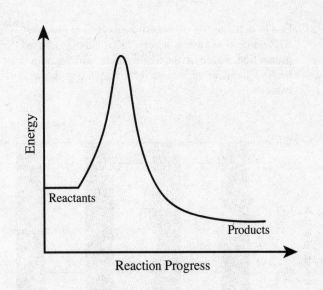

20. If the reaction shown above were catalyzed by an enzyme

(A) The reaction would become exergonic
(B) The reaction would remain endergonic
(C) The activation energy would increase
(D) The rate of reaction would increase

21. Fast twitch muscles do not utilize oxygen to produce ATP, while slow twitch muscles rely on oxygen for ATP production. Based on the figure below, which group is LEAST likely to utilize anaerobic respiration during exercise?

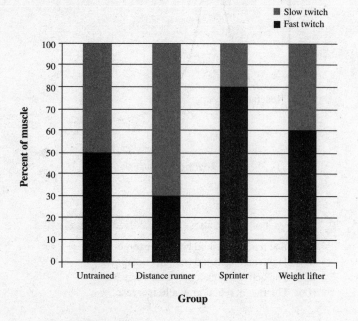

(A) Untrained
(B) Distance runner
(C) Sprinter
(D) Weight lifter

22. The rate of a reaction catalyzed by an enzyme would NOT be affected by

(A) pH
(B) temperature
(C) amount of enzyme present
(D) the number of times the enzyme has previously been used

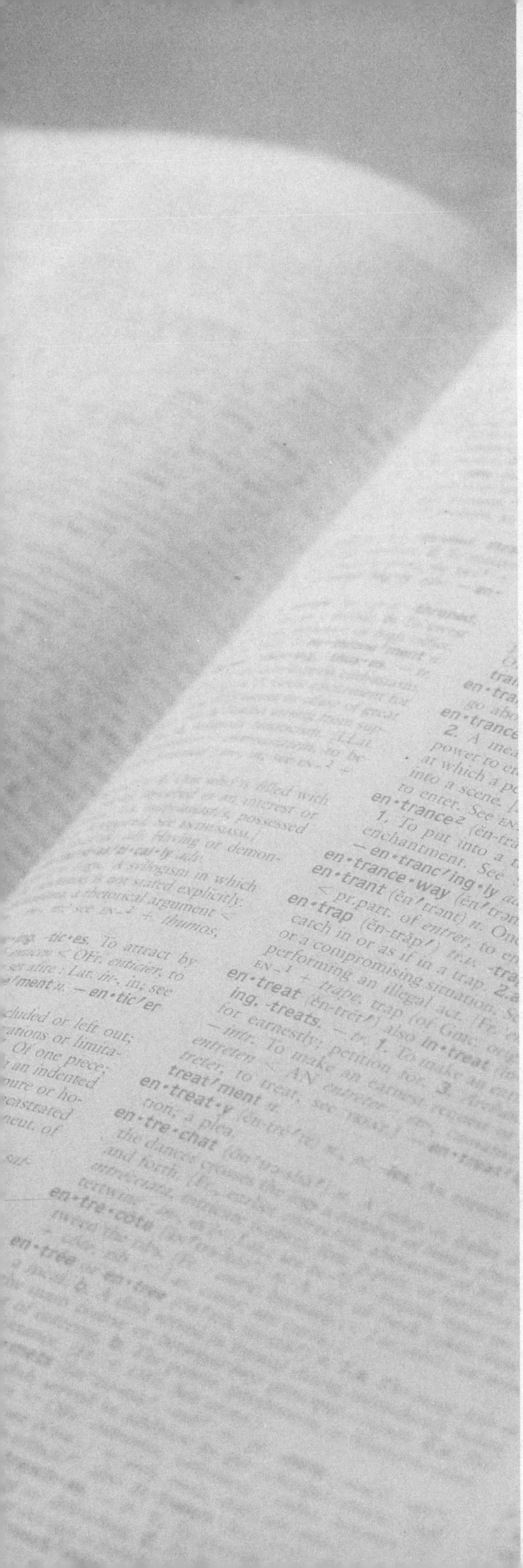

Chapter 6 Big Idea 2 Drill 1 Answers and Explanations

ANSWER KEY

1. C
2. B
3. C
4. C
5. B
6. A
7. C
8. A
9. A
10. D
11. B
12. C
13. B
14. A
15. C
16. D
17. D
18. A
19. B
20. D
21. B
22. D

ANSWERS AND EXPLANATIONS

Multiple-Choice Questions

1. **C** According to the chart, the basic unit for malt sugar is a disaccharide, while the basic unit for corn syrup is a monosaccharide. Disaccharides are split into monosaccharides via hydrolysis. Condensation and dehydration synthesis are both terms for the linkage of sugar molecules in which a water molecule is lost so eliminate (A) and (B). Hydrogen bonding is not present in carbohydrates, so cross off (D).

2. **B** Peptide bonds are the linkage between amino acids to form a polypeptide, which after folding will become a three-dimensional protein. A peptide bond is formed when the carboxyl group of one amino acid joins with the amino group of another amino acid. I represents an amino functional group, and III represents a carboxyl functional group. II represents a methyl group, which does not play a role in peptide bonds, (A), (C), and (D).

3. **C** Transmembrane proteins are proteins that sit in the plasma membrane of a cell and have both intracellular and extracellular regions. The membrane-spanning region is the part within the lipid bilayer. Because the lipid bilayer is made up of phospholipids, with hydrophobic lipid tails in the middle, you can expect that the membrane-spanning portion of a protein would also be hydrophobic. According to the chart, arginine is hydrophilic, so eliminate choices (B) and (D). Isoleucine is very hydrophobic, and alanine is hydrophobic, so both could be present in the membrane-spanning portion of a transmembrane protein.

4. **C** Mitochondria are known as the powerhouses of the cell, and this organelle's main role is ATP production. Thus, cells that require a large amount of energy would have a large amount of mitochondria. Fat, hair, and skin cells require less energy than heart muscle cells, which need to contract in order for the heart to pump, eliminating all other choices except (C).

5. **B** Both glycolysis and oxidative phosphorylation result in the net production of ATP molecules. While glycolysis requires two molecules of ATP (and then produces two molecules of ATP, for a net production of two ATP), oxidative phosphorylation does not require any input of ATP (A). Oxidative phosphorylation requires oxygen as the final electron acceptor, but glycolysis can occur in anaerobic conditions (C). While oxidative phosphorylation occurs in across the inner mitochondrial membrane, glycolysis takes place in the cytoplasm (D).

6. **A** PEP carboxylase, the enzyme that fixes phosphoenolpyruvate (PEP) in the C_4 pathway, has a very high affinity for carbon dioxide, allowing it to combine CO_2 with PEP to form oxaloacetate even in conditions when there is very low CO_2, such as in dry climates. Eliminate choice (B) as it says the opposite. RuBP carboxylase is used in the C_3 pathway, so eliminate choices (C) and (D).

7. **C** Ligase is used to bring together Okazaki fragments on the lagging strand of DNA; because the leading strand is synthesized in a continuous fashion, there is no need for ligase. Helicase unwinds the parent DNA double-helix into two strands to allow DNA replication to begin (A). Polymerase is responsible for adding nucleotides to the growing daughter DNA—both leading and lagging strands (B). RNA primase catalyzes the synthesis of RNA primers, which must be put down in order to start the synthesis of both leading and lagging DNA strands (D).

8. **A** To make mRNA from the DNA gene, RNA polymerase must first be used to synthesize the mRNA strand in the process called transcription. The mRNA then must be processed to remove introns, which is accomplished via the spliceosome. Finally, a ribosome must be present for translation—the synthesis of a polypeptide chain from an mRNA. While tRNAs are used in the process of translation, tRNA is not an enzyme so that eliminates (C) and (D).

9. **A** Because DNA polymerase must be available during S phase of mitosis, it must be created during G1 stage of interphase. G1 is a growing stage, and the cell gets ready for mitosis by producing all of the enzymes it will need for DNA replication, including DNA polymerase. G2 stage and prophase are after S stage, so DNA polymerase would not be available in S phase if this enzyme were made then. All proteins are synthesized during the cell cycle because this is an ongoing cycle that includes the entire life of the cell (D)!

10. **D** Mitosis is the simple replication of hereditary material followed by cellular division, resulting in two identical daughter cells that are also identical to the parent cell. This process allows for asexual reproduction, growth of an organism, and repair of damaged cells in an organism. Mitosis does not play a role in sexual reproduction, which involves two haploid gametes from two individual parents coming together to form a new diploid cell.

11. **B** The figure shows homologous chromosomes, or tetrads, lining up in the middle of the cell, as occurs in Metaphase I of meiosis. While you could argue that the tetrads may be starting to separate as would occur in Anaphase I of meiosis, there is no separation of sister chromatids at the centromere in Anaphase I—this occurs in mitotic anaphase (A). In mitosis, duplicated chromosomes (two sister chromatids joined by a centromere) line up single-file down the center of the cell (C). G2 phase of interphase would be characterized by chromosomes composed of sister chromatids, but they would not be lined up as in the figure (D).

12. **C** The small intestine is the organ most responsible for the digestion of food into basic units via enzymes released into this part of the gastrointestinal tract. The main function of the esophagus is simply to move food from the mouth into the stomach via peristalsis (A). While many people think of the stomach as digesting food, its main job is actually to store food; it only temporarily digests proteins (B). The large intestine reabsorbs water and salts and stores feces (D).

13. **B** One of the main features of the stomach is its acidity, which allows the enzyme trypsin to function properly and also kills bacteria. When a food bolus moves from the stomach to the small intestine, some of these acidic stomach juices move with it. While the lining of the stomach is designed to

withstand this acidity, the lining of the small intestine is not, and thus it needs special protection. Basic pancreatic bicarbonate workers as a buffer and neutralizes the acidic chyme that enters the small intestine. Bicarbonate is a base and thus would not increase acidity in the small intestine (D). While bicarbonate does play a role in regulating blood pH, this is a function of bicarbonate ions in the kidney and not the pancreas (A). Bicarbonate is formed from a reaction of carbon dioxide and water; it is not a waste product from protein breakdown (C).

14. **A** Oxygenated blood returns to the heart after oxygenation in the lungs—it enters the left atrium through the pulmonary veins, then travels from the left atrium to the left ventricle through the left atrioventricular (bicuspid) valve, and then finally exits the heart into the aorta via the aortic semi-lunar valve. Deoxygenated blood is returned to the heart via the vena cava and dumped into the right atrium, then travels to the right ventricle via the right atrioventricular (tricuspid) valve, and exits the heart through the pulmonary semilunar valve to enter the pulmonary artery and travel to the lungs. There it will be oxygenated and then return to and travel through the left side of the heart, as explained above.

15. **C** Peptide hormones cannot diffuse across the cell membrane; instead, they bind to extracellular receptors which trigger second messenger systems within the cell. This is a relatively fast process as compared to steroid hormones that diffuse across the cell membrane, enter the nucleus, and then bind to DNA to affect transcription (A). Steroid hormones do not bind extracellular receptors but instead bind to DNA in the nucleus of a cell (B). Peptide hormones do not bind to DNA and cannot even enter a cell, let alone the nucleus of a cell (D).

16. **D** After an egg is fertilized by a sperm, the new zygote quickly divides to undergo a series of changes. First, the zygote becomes a morula, a solid ball of cells. Next, the morula becomes a blastula, a hollow ball of cells. Finally, the blastula becomes a gastrula, in which three germ layers differentiate.

17. **D** The yolk sac provides food for the growing embryo and thus is its source of energy. The amnion forms a fluid-filled sac which helps to protect the delicate embryo (A). The chorion is the outermost extraembryonic membrane, surrounding all of the others, and later will transfer nutrients from the mother to the fetus (but not to the embryo because it is too early) (B). The mesoderm is one of the three germ layers present in the gastrula and will eventually give rise to many body structures such as bones, blood, and muscles.

18. **A** Heterotrophs must get their energy from other organisms, but this could be from autotrophs OR heterotrophs. Autotrophs use basic building blocks plus light energy to create chemical energy via photosynthesis, (B) and (D). Heterotrophs do break down organisms that they consume into chemical energy (C).

19. **B** ATP is one of the products of glycolysis, so it makes sense that it would function as an inhibitor on one of the enzymes necessary to glycolysis. If there is a surplus of ATP present, there would be no need to go through glycolysis. ADP is the molecule that is phosphorylated to make ATP; it is necessary for glycolysis and should not inhibit the process (A). Fructose-6-phosphate is the molecule

phosphorylated by phosphofructokinase, so it should not inhibit the enzyme (C). Likewise, glucose is the starting point for glycolysis, so it would not make sense for glucose to inhibit one of the enzymes needed for glycolysis (D).

20. **D** An enzyme that catalyzes a reaction will lower the activation energy for the reaction and thus increase the rate of the reaction (C). Enzymes do not change the thermodynamics of a reaction, only the kinetics, and thus could not change a reaction to become exergonic (A). Furthermore, the reaction shown is exergonic, not endergonic (B).

21. **B** Distance runners have the lowest percentage of slow twitch muscle and are more likely to utilize their overwhelming number of slow twitch muscle for aerobic respiration during exercise. All three other groups have more fast twitch muscle than the distance runners and thus would be more likely to utilize anaerobic respiration in their fast twitch muscles than the distance runners.

22. **D** An enzyme increases the rate of a reaction without undergoing any changes to itself; it is completely preserved at the end of the reaction, so it does not matter how many times an enzyme has already been used to catalyze a reaction. It will still function the same way in increasing the rate of reaction. pH and temperature both have effects on enzymatic function (A) and (B). If more enzyme is added to a reaction, the rate of reaction will increase because their will be more enzyme available to interact with substrate (C).

Chapter 7
Big Idea 3 Drill 1

BIG IDEA 3 DRILL 1

Multiple-Choice Questions

1. Which difference between DNA and RNA accounts for the difference in their names?

    (A) DNA is double-stranded, while RNA is single stranded.
    (B) The sugar molecule in DNA has one less hydroxyl group than the sugar molecule in RNA.
    (C) DNA uses thiamine as a base, while RNA uses uracil as a base.
    (D) DNA is the hereditary material, while RNA is utilized in protein synthesis.

2. Which of the following types of proteins does NOT transmit materials or information from the extracellular environment to the intracellular environment?

    (A) Adhesion proteins
    (B) Channel proteins
    (C) Receptor proteins
    (D) Recognition proteins

3. Which of the following cell parts are common to both plant and animal cells?

    I.   Cell wall
    II. Centrioles
    III. Ribosomes

    (A) I only
    (B) I and II only
    (C) II and III only
    (D) III only

4. Which type of cell junction facilitates communication between adjacent cells?

    (A) Desmosome
    (B) Gap junction
    (C) Tight junction
    (D) Clathrin

5. Oxidative phosphorylation utilizes the energy held in electron carriers (such as NADH) to product ATP. Electron carriers from which process would NOT need to be transported in the cell prior to entering the electron transport chain?

   (A) Fermentation
   (B) Glycolysis
   (C) Krebs cycle
   (D) Anaerobic respiration

6. Which of the following are responsible for control of gas exchange and transpiration in plants?

   (A) Guard cells
   (B) Stomates
   (C) Thylakoids
   (D) Vascular bundles

7. When genes in a chromosome are active and available for transcription, this DNA is

   (A) called euchromatin, and it is loose
   (B) called euchromatin, and it is condensed
   (C) called heterochromatin, and it is loose
   (D) called heterochromatin, and it is condensed

8. Which is true concerning the importance in consistency of complementary base-pairing between each strand of a DNA double-helix?

   (A) Purines pair with purines, and pyrimidines pair with pyrimidines so that the area taken up by bases is consistent down the ladder.
   (B) Purines pair with pyrimidines so that the area taken up by bases is consistent down the ladder.
   (C) Purines pair with purines, and pyrimidines pair with pyrimidines so that the same number of hydrogen bonds between bases is consistent down the ladder.
   (D) Purines pair with pyrimidines so that the same number of hydrogen bonds between bases is consistent down the ladder.

9. RNA

   (A) is the hereditary blueprint of the cell
   (B) is replicated and passed on from parent to daughter cells
   (C) is an intermediary molecule that carries the instructions in DNA
   (D) is the final structure in the flow of genetic information

10. Translation of proteins requires a lot of energy, so cells only make proteins that are necessary for their specific function at any given time. How is the production of proteins primarily controlled?

   (A) At the level of transcription, by transcription factors which bind to DNA
   (B) At the level of transcription, by primases which bind to RNA
   (C) At the level of translation, by transcription factors which bind to DNA
   (D) At the level of translation, by primases which bind to RNA

11. The genetic material of retroviruses is

   (A) made of DNA that can be inserted into the host genome
   (B) made of RNA that must be converted to DNA to enter the host genome
   (C) made of DNA that must be converted to RNA to enter the host genome
   (D) very stable from one generation to the next because the enzymes involved in retrovirus replication of hereditary material rarely make mistakes

12.. The frequency of crossing over between two linked alleles is

   (A) directly proportional to the distance between them
   (B) inversely proportional to the distance between them
   (C) completely random
   (D) impossible to determine

13. What might trigger the release of bile from the gall bladder into the small intestine?

(A) Consuming a healthy vegetable-based meal
(B) Consuming a highly fatty meal
(C) Fasting for 12 hours
(D) Liver failure

14. Which is the best explanation for control of human respiratory rate?

(A) Chemoreceptors detect blood pH and send signals to muscles to increase respiration rate if blood pH is too low.
(B) Chemoreceptors detect blood pH and send signals to muscles to increase respiration rate if blood pH is too high.
(C) Mechanoreceptors detect oxygen levels in the lungs and send signals to muscles to increase respiration rate if alveolar oxygen concentration is too low.
(D) Mechanoreceptors detect oxygen levels in the lungs and send signals to muscles to increase respiration rate if alveolar oxygen concentration is too high.

15. Which cell surface protein communicates with T cells to differentiate between self and non-self material?

(A) Clathrin
(B) G-protein
(C) Major histocompatibility complex
(D) Sodium-potassium ATPase

16. Which is/are true regarding hormonal regulation of the kidney?

   I. ADH controls the volume of urine.
  II. ADH levels go up in response to dehydration.
 III. Aldosterone regulates potassium reabsorption at the proximal convoluted tubule.

(A) I only
(B) I and II only
(C) II and III only
(D) I, II, and III

17. In the diagram below, which of the following is NOT true of the neuron labeled "2"?

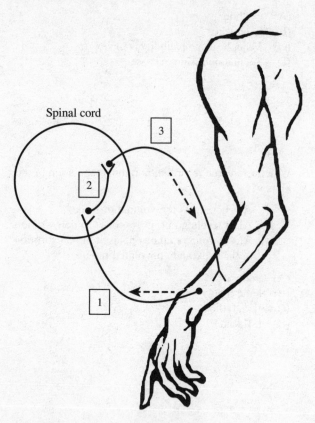

(A) It allows information gathered by the sensory nervous system to result in muscle movement.
(B) It is called an interneuron.
(C) It is contained in the central nervous system.
(D) It transmits information from a motor neuron to a sensory neuron.

18. Which of the following in the diagram is responsible for creating the concentration gradient that allows for the rush of ions that begins an action potential in a neuron?

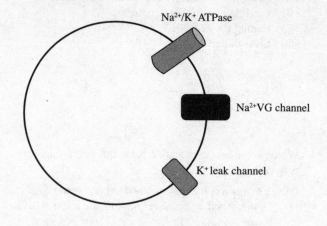

(A) The Na^{2+}/K^+ ATPase pumping Na^{2+} ions into the cell
(B) The Na^{2+}/K^+ ATPase pumping Na^{2+} ions out of the cell
(C) The K^+ leak channel allowing K^+ ions to leave the cell
(D) The K^+ leak channel allowing K^+ ions to enter the cell

19. Which of the following hormones released from the pituitary triggers ovulation during the menstrual cycle?

(A) Luteinizing hormone
(B) Follicle-stimulating hormone
(C) Estrogen
(D) Progesterone

20. Based on the graph of different hormone levels during the menstrual cycle shown below, which is LEAST likely to be true?

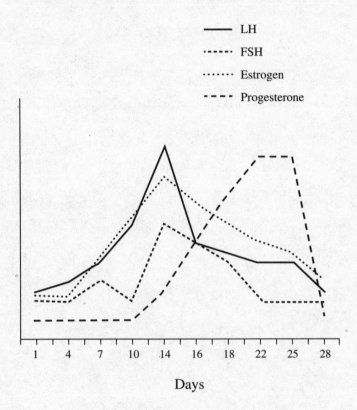

——— LH

----- FSH

········· Estrogen

– – – Progesterone

Days

(A) Ovulation triggers an increase in progesterone.
(B) The drop in estrogen causes the flow phase of the menstrual cycle.
(C) The rise in progesterone causes a rise in estrogen.
(D) The rise in LH causes a rise in estrogen.

21. How do prokaryotes ensure continued genetic variation in populations?

(A) Vertically via mitosis
(B) Vertically via meiosis
(C) Vertically via binary fission
(D) Horizontally via transduction, transformation, and conjugation

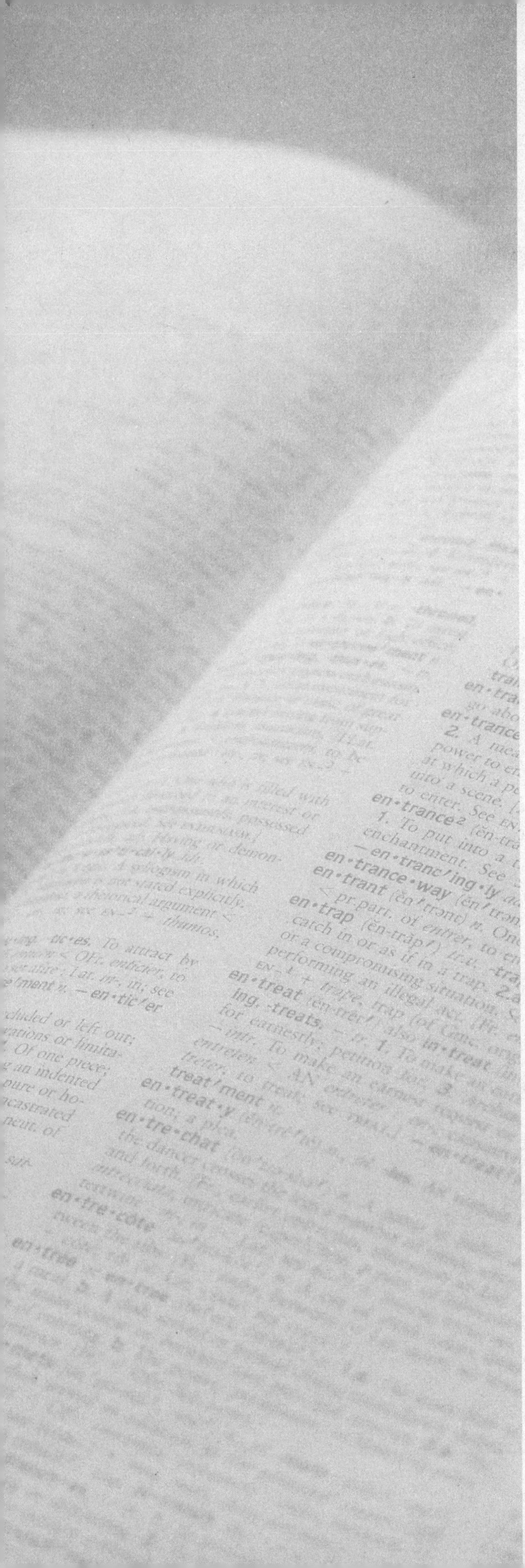

Chapter 8
Big Idea 3
Drill 1 Answers and Explanations

ANSWER KEY

1. B
2. A
3. D
4. B
5. C
6. A
7. A
8. B
9. C
10. A
11. B
12. A
13. B
14. A
15. C
16. B
17. D
18. B
19. A
20. C
21. D

ANSWERS AND EXPLANATIONS

Multiple-Choice Questions

1. **B** Deoxyribonucleic acid (DNA) differs from ribonucleic acid (RNA) in several ways, but as the prefix "deoxy-" implies, the ribose sugar in DNA has lost an oxygen as compared with the ribose sugar in RNA. DNA is double-stranded and RNA is typically single-stranded, but this is not related to their names (A). DNA does have thymine as a base while RNA has uracil, but again, this difference does not enter into the naming scheme (C). DNA is the hereditary material, and RNA is necessary for protein synthesis (translation), but this difference is also not a part of the nomenclature (D).

2. **A** Adhesion proteins form junctions between adjacent cells and so do not play a roll in providing information about the extracellular environment to the inside of the cell or transmitting materials into the cell. Channel proteins allow specific molecules or ions to enter and exit the cell, so eliminate choice (B). Receptor proteins serve as docking sites for hormones and other extracellular proteins, which then typically causes an intracellular change in function in response to the receptor being filled; eliminate choice (C). Recognition proteins are found on the extracellular surface, and provide information to surrounding cells about their cell type, which is valuable information about the surrounding environment. Eliminate choice (D).

3. **D** Ribosomes are found in all types of cells (prokaryotic, plant, and animal cells). A cell wall (I) is found surrounding plant cells, but not animal cells, so eliminate choices (A) and (B). Centrioles (II) are not found in plant cells, so eliminate choice (C).

4. **B** Gap junctions form channels between cell membranes to allow communication and transfer of materials between the cytoplasm of two adjacent cells. Desmosomes hold adjacent animal cells very tightly together (A). Tight junctions seal off the space between two adjacent cells so that no extracellular material can get between cells (C). Clathrin coats pits that are utilized for certain types of endocytosis (D).

5. **C** The Krebs cycle takes place in the mitochondrial matrix, so any electron carriers made in the process are immediately available to enter the electron transport chain along the mitochondrial inner matrix. Fermentation occurs in the cytosol as part of anaerobic respiration, so it would not be able to use the electron transport chain to produce more energy (A). Glycolysis takes place in the cytosol, so any electron carriers produced here would need to be transported to the mitochondria. Anaerobic respiration does not occur in the mitochondria but in the cytosol, so again any electron carriers from this process would need transport to the mitochondria.

6. **A** Guard cells surround stomates, controlling the stomates' opening and closing. Stomates are the cells that facilitate gas exchange and transpiration through the lower epidermis of a plant, but it is the guard cells which provide control (A). Thylakoids are the cells making up the grana and contain chlorophyll as well as enzymes required for photosynthesis (C). The vascular bundles, particularly xylem and phloem, transport materials through a plant (D).

7. **A** This is an example of a two-by-two question; you need to know if the DNA of active genes available for transcription is condensed or loose, and you must know the name for this type of configuration. In order for RNA polymerase to begin access the DNA to begin transcription, the DNA must be loose, so eliminate choices (B) and (D). Loose DNA is known as euchromatin; condensed DNA is called heterochromatin (C).

8. **B** Complimentary bases in DNA pair in a predictable manner—adenine (a purine) always pairs with thymine (a pyrimidine), and guanine (a purine) always pairs with cytosine (a pyrimidine). Thus, purines pair with pyrimidines, so eliminate choices (A) and (C). The number of hydrogen bonds between bases is NOT consistent—there are three hydrogen bonds between G and C and two hydrogen bonds between A and T—so eliminate choice (D). Because purines are larger structures than pyrimidines, this consistent pairing of purine with pyrimidine ensures that the space taken up by base pairs on the inside of the DNA double helix is consistent.

9. **C** RNA is the intermediary molecule between DNA and protein in the central dogma of molecular biology (DNA → RNA → Protein), and it carries the information in DNA to the ribosome to allow protein synthesis. DNA makes up the hereditary blueprint of the cell (A), and DNA is replicated and passed on from parent to daughter cells (B). The final structure in the flow of genetic information is protein (D).

10. **A** Protein production is regulated at the level of transcription by appropriately named transcription factors (repressors or enhancers) that bind to DNA to either increase or decrease transcription. Translation is not regulated by the cell—once an RNA molecule is made, a protein will also be made through translation. Eliminate choices (C) and (D). Primases do not bind RNA and do not play a role in transcription regulation; primases are used in DNA replication to lay down RNA primer prior to DNA polymerase adding nucleotides to synthesize daughter DNA, so you can eliminate (B).

11. **B** Retroviruses contain RNA that must be reversed transcribed into DNA in order for the virus to enter the lysogenic cycle and insert its DNA into the host genome. Interestingly, this violates the central dogma of molecular biology (DNA → RNA → Protein), providing another example of the strange nature of viruses. The genetic material of retroviruses is not DNA—if these viruses had a DNA genome, they could simply insert their material into the host genome without going through reverse transcription. The RNA in retroviruses is very unstable and often goes through mutation because the enzymes involved in replication of the RNA are not very precise and often make mistakes.

12. **A** Crossing over is more likely to occur the further apart two linked genes are, so the distance between then is directly proportional to their frequency of crossing over. An inverse proportion means the opposite would be true—the further apart genes are, the less they would cross over, which is the opposite of reality (B). Frequency of crossing over is not completely random and can be determined by looking at genes frequency in offspring, (C) and (D).

13. **B** Bile is made in the liver, stored in the gall bladder, and then released into the small intestine to emulsify fats (break down fats into smaller droplets for easier digestion) so it makes sense that it would be released in response to consumption of a fatty meal. A healthy vegetable-based meal or fasting should not trigger the release of bile, which is only needed for fat emulsification. Because the liver produces bile, liver failure could lead to problems with bile production but should not have an effect on the release of bile from the gall bladder (D).

14. **A** Chemoreceptors are able to detect pH of the blood, and if blood pH is too low (acidic), send signals to respiratory muscles in order to increase respiration rate to blow off carbon dioxide (an acid) to lower the pH of blood. If blood pH is too low, the opposite would happen, and the respiration rate would slow down (B). Mechanoreceptors detect mechanical pressure or distortion and are typically found on the skin, (C) and (D).

15. **C** The major histocompatibility complex proteins display cell contents on the outside of the cell, and T cells look at what the major histocompatibility complex proteins are holding and can determine if its contents are self or non-self. This allows the T cells to decide if cells are infected or cancerous and launch an immune attack on abnormal (non-self) cells. Clathrin coats pits that are utilized for certain types of endocytosis; it does not play a role in the immune system (A). G-proteins are second messengers that allow extracellular protein receptors on the cell membrane to transmit a signal to the inside of the cell (B). The sodium-potassium ATPase is an active transport protein that uses ATP to transport sodium out of the cell and potassium into the cell (D).

16. **B** ADH controls the volume of urine by increasing the volume of urine when a person is well-hydrated and decreasing the volume or urine (and thus saving more water inside the body) when a person is dehydrated. The more ADH present, the more water is saved inside the body, so it makes sense that ADH levels would go up if a person were dehydrated. Aldosterone regulates sodium reabsorption and potassium secretion at the distal convoluted tubule in the nephron (III).

17. **D** The neuron labeled "2" is called an interneuron which transmits information from a sensory neuron (1) to a motor neuron (3) and thus allows information from the senses to cause muscle movement, (A) and (B). It is completely within the spinal cord, a component of the central nervous system (C).

18. **B** The sodium-potassium ATPase pumps sodium ions out of the cell and so creates the concentration gradient of much more sodium outside of the cell than inside. So when the voltage-gated sodium channels open, sodium ions rush into the cell, which results in the large and swift increase in membrane potential known as an action potential. The sodium-potassium ATPase pumps sodium

ions out of the cell, not into the cell. The potassium leak channel allows potassium ions to move in either direction, but because the concentration of potassium is much higher inside of the cell, the potassium ions leak out of the cell; regardless, the potassium concentration is not responsible for the sodium concentration gradient which results in the rush of sodium ions into the cell that begins an action potential.

19. **A** A spike in luteinizing hormone, released by the anterior pituitary, triggers ovulation around day 14 of the menstrual cycle. Follicle-stimulating hormone stimulates follicular maturation in females in preparation for ovulation (B). Estrogen and progesterone are produced by the ovaries; estrogen promotes female secondary sex characteristics and causes the endometrial lining of the uterus to thicken, while progesterone is responsible for maintenance of the endometrial lining prior to menstruation.

20. **C** Looking at the graph of different hormonal levels during the menstrual cycle, the rise in progesterone during the second half of the cycle accompanied by a drop in estrogen levels, so it is unlikely that a rise in progesterone causes a rise in estrogen. Ovulation, which occurs in response to the LH spike around day 14, does result in an increase in progesterone levels, which start to rise around day 14 (A). Estrogen begins to drop following ovulation and continues to fall through day 28, so it makes sense that this drop could lead to menstruation, or the flow phase, at the end of the cycle (B). The rise in LH during the first half of the cycle is mimicked by the similar rise in estrogen, so it is possible that the rise in LH could be causing the rise in estrogen, directly or indirectly (D).

21. **D** Transduction, transformation, and conjugation are three different methods by which bacteria are able to take up new DNA or share DNA between organisms. Reproduction via binary fission, a vertical process, does not lead to genetic variation in prokaryotes (although sexual reproduction does in eukaryotes), so eliminate choice (C). Beyond this, prokaryotes do not go through mitosis or meiosis, so eliminate choices (A) and (B).

Chapter 9
Big Idea 4 Drill 1

BIG IDEA 4 DRILL 1

Multiple-Choice Questions

1. How is carbon dioxide primarily transported in the body?

   (A) On hemoglobin
   (B) Dissolved in plasma
   (C) As hydrogen ions
   (D) As bicarbonate ions

2. The heterotroph hypothesis proposes all of the following EXCEPT

   (A) early simple cells used organic molecules as sources of food
   (B) complex cells evolved from simple cells
   (C) early heterotrophs and early autotrophs came from completely separate cellular ancestors
   (D) early autotrophs produced the oxygen that is part of the Earth's atmosphere

3. An example of protein secondary structure is

   (A) hydrogen bonds
   (B) a linear chain of amino acids
   (C) sulfide bonds
   (D) alpha helices

4. Spermatogenesis and oogenesis differ in that

   (A) spermatogenesis begins in the male fetus, while oogenesis does not begin until female puberty
   (B) spermatogenesis ends around age 70 for males, while oogenesis continues indefinitely in females
   (C) spermatogenesis results in one sperm cell for each diploid cell, while oogenesis results in four ova per diploid cell
   (D) spermatogenesis results in four sperm cells for each diploid cell, while oogenesis results in one ovum per diploid cell

5. Based on the graph below, it is reasonable to assume that

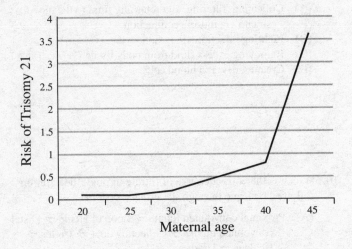

(A) risk of non-disjunction decreases as mothers' get older

(B) risk of non-disjunction increases as mothers' get older

(C) risk of base mutations decreases as mothers' get older

(D) risk of base mutations increases as mothers' get older

6. Human blood type is the most common example of codominance. Which of the following explanations for blood type codominance is most likely true?

(A) The A allele is dominant over the B allele, so only the A protein will be expressed in a person with both alleles.

(B) The B allele is dominant over the A allele, so only the B protein will be expressed in a person with both alleles.

(C) The A allele and B allele cancel each other out, so neither the A or B protein will be expressed in a person with both alleles.

(D) The A allele and B allele are both expressed independently of each other, so both the A protein and B protein will be expressed in a person with both alleles.

7. The pancreas has both endocrine functions and exocrine functions. The endocrine pancreas

(A) releases hormones into the bloodstream, while the exocrine pancreas releases enzymes into the digestive tract

(B) releases hormones into the digestive tract, while the exocrine pancreas releases enzymes into the bloodstream

(C) releases enzymes into the bloodstream, while the exocrine pancreas releases hormones into the digestive tract

(D) releases enzymes into the digestive tract, while the exocrine pancreas releases hormones into the bloodstream

8. The figure below depicts the hemoglobin saturation curve, which indicates how much hemoglobin is carrying oxygen in different conditions. The solid black line shows normal conditions, while the dotted line and the conditions listed on the right represent exercise.

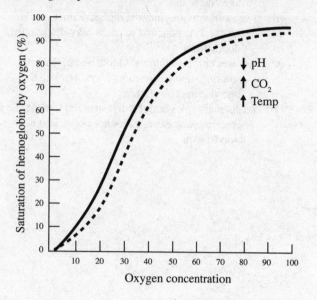

From the graph, what can you conclude about oxygen delivery to muscles during exercise?

(A) More oxygen is delivered to muscles during exercise because the saturation of hemoglobin by oxygen increases.

(B) More oxygen is delivered to muscles during exercise because the saturation of hemoglobin by oxygen decreases.

(C) Less oxygen is delivered to muscles during exercise because the concentration of carbon dioxide increases.

(D) Less oxygen is delivered to muscles during exercise because the concentration of carbon dioxide decreases.

9. Which is NOT a function of the immune system?

(A) Collecting, filtering, and returning fluid to the blood stream via muscle contraction

(B) Fighting infection via lymphocytes

(C) Removing excess fluid from body tissue

(D) Creating new red blood cells

10. Which is the correct course of filtrate through the nephron followed by urine through the urinary system?

(A) Proximal convoluted tubule → Loop of Henle → Distal convoluted tubule → Collecting duct → Ureter → Bladder → Urethra

(B) Distal convoluted tubule → Loop of Henle → Proximal convoluted tubule → Collecting duct → Urethra → Bladder → Ureter

(C) Proximal convoluted tubule → Distal convoluted tubule → Loop of Henle → Collecting duct → Ureter → Bladder → Urethra

(D) Proximal convoluted tubule → Distal convoluted tubule → Loop of Henle → Collecting duct → Bladder → Urethra

11. In the process of making urine, reabsorption and secretion

    I. Occur in the tubules of the nephron
   II. Are opposite processes
  III. Do not occur if blood pH is normal

(A) I only
(B) I and II
(C) II and III
(D) I, II, and III

12. The layer of skin containing the structures that secrete water and ions is the

(A) dermis
(B) epidermis
(C) stratum corneum
(D) subcutaneous tissue

13. Which is NOT true of the myelin sheaths surrounding an axon?

(A) They insulate the axon, which allows impulses to travel faster.
(B) They do not cover nodes of Ranvier.
(C) They facilitate saltatory conduction.
(D) They create a domino effect, whereby every part of the axon cell membrane must be individually depolarized.

14. Acetylcholinesterase

(A) is released from the end of an axon when calcium enters the terminal end of the axon
(B) is picked up very quickly by the dendrites on the post-synaptic neuron
(C) can stimulate muscles to contract
(D) breaks down neurotransmitters in the synaptic clefts of the parasympathetic nervous system

15. Which part of the nervous system is responsible for controlling skeletal muscle movement?

(A) Autonomic nervous system
(B) Parasympathetic nervous system
(C) Somatic nervous system
(D) Sympathetic nervous system

16. Why is it more difficult for an action potential to be fired immediately following a previous action potential?

(A) The cell membrane potential is too positive, making it hard to reach threshold.
(B) The cell membrane potential is too negative, making it hard to reach threshold.
(C) The sodium channels are inactive and cannot open.
(D) The potassium channels are inactive and cannot open.

17. Which is an accurate statement describing fertilization and implantation?

(A) Both fertilization and implantation usually occur in the uterus.
(B) Fertilization usually occurs in the ovary, while implantation occurs in the fallopian tube.
(C) Fertilization usually occurs in the fallopian tube, while implantation occurs in the uterus.
(D) Fertilization usually occurs in the ovary, while implantation occurs in the uterus.

18. This poster below created by the National Organization on Fetal Alcohol Syndrome indicates that the most common site of birth defects due to alcohol following the embryonic period is

Fetal Development Chart

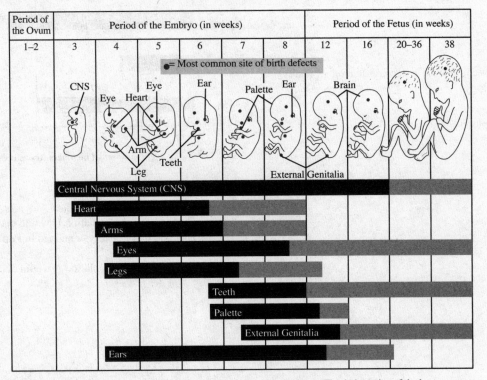

Vulnerability of the fetus to defects during different periods of development. The dark portion of the bars represents the most sensitive periods of development, during which teratogenic effects on the sites listed would result in major structural abnormalities in the child. The lighter portion of the bars represents periods of development during which physiological defects and minor structural abnormalities would occur.

SOURCE: Adapted from Moore 1993.

(A) The ears
(B) The eyes
(C) The brain
(D) The heart

19. Where would you find stunted conifers, caribou, wolves, and moose?

(A) Grasslands
(B) Taiga
(C) Temperate deciduous forest
(D) Tundra

20. Which of the following is NOT a density-dependent factor on population ecology?

(A) Hurricanes
(B) Resource depletion
(C) Competition
(D) Predation

21. Based on the graph of four different populations sharing an ecosystem, what would you expect to happen if the numbers of members of population B suddenly increased by 25%?

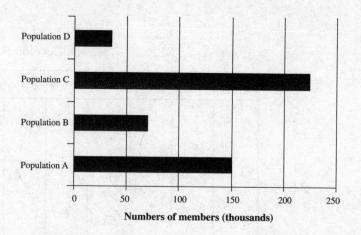

Numbers of members (thousands)

(A) The amount of energy contained in Population D would decrease
(B) The biomass of Population D would decrease
(C) The amount of energy contained in Population A would increase
(D) The biomass of Population A would decrease

22. B cells played a critical role in mediating the humoral adaptive immune response to foreign antigens. Which of the following is NOT a characteristic of B cells?

(A) They are derived from bone marrow.
(B) They produce antibodies.
(C) They release chemicals to destroy infected cells.
(D) They may provide memory to an infection.

23. The glomerulus initially filters many components out of the blood. All of the following would be expected to be filtered out of the blood EXCEPT

(A) urea
(B) water
(C) proteins
(D) salts

24. Alcohol is an antidiuretic hormone (ADH) inhibitor. Which of the following accurately describes how consumption of alcohol would affect water retention and the osmolarity of the blood?

(A) Alcohol consumption will result in increased water retention and increased blood osmolarity.
(B) Alcohol consumption will result in increased water retention and decreased blood osmolarity.
(C) Alcohol consumption will result in decreased water retention and decreased blood osmolarity.
(D) Alcohol consumption will result in decreased water retention and increased blood osmolarity.

25. Which of the following accurately lists the order by which filtrate is processed in the nephron?

(A) Distal Convoluted Tubule, Loop of Henle, Proximal Convoluted Tubule, and Collecting Duct
(B) Proximal Convoluted Tubule, Loop of Henle, Distal Convoluted Tubule, and Collecting Duct
(C) Proximal Convoluted Tubule, Distal Convoluted Tubule, Loop of Henle, and Collecting Duct
(D) Collecting Duct, Proximal Convoluted Tubule, Loop of Henle, Distal Convoluted Tubule

Chapter 10
Big Idea 4
Drill 1 Answers and Explanations

ANSWER KEY

1. D
2. C
3. D
4. D
5. B
6. D
7. A
8. A
9. D
10. A
11. B
12. A
13. D
14. D
15. C
16. B
17. C
18. C
19. B
20. A
21. D
22. C
23. C
24. D
25. B

ANSWERS AND EXPLANATIONS

Multiple-Choice Questions

1. **D** Most carbon dioxide travels through the body by entering red blood cells and combining with water to form bicarbonate ions (HCO_3^-). Carbon dioxide can be carried on hemoglobin, but this accounts for about 20% of carbon dioxide, whereas more than 70% of carbon dioxide is transported as bicarbonate ions (A). A small amount of carbon dioxide can dissolve in plasma, but again, this is not the primary mode for transport (B). Hydrogen ions are present in the body but are a separate entity from carbon dioxide (C).

2. **C** The heterotroph hypothesis proposes that all organisms originally came from the same simple cells which evolved into complex cells; the earliest organisms were heterotrophs which used organic molecules as food, and then later on some organisms found a way to make their own food, becoming autotrophs, eliminating (A) and (B). The autotrophs used energy from the sun to make their own food, and one byproduct of this process was oxygen, which they released into the atmosphere (D).

3. **D** Alpha helices and beta-pleated sheets are two examples of secondary protein structures—both are due to twisting of the polypeptide chain. The linear chain of amino acids is known as the protein's primary structure (A). Sulfide bonds contribute to proteins' tertiary structure (C). Hydrogen bonds do not play an important role in higher order protein structure (A).

4. **D** Gametogenesis is slightly different in men and women with one major difference being that spermatogenesis results in four sperm cells, while oogenesis only results in one ovum. This is due to the formation of one polar body, a tiny cell that degenerates, during each of Meiosis I and Meiosis II. Because the egg supplies all of the organelles for the zygote, it is important that the female gamete be large and have a lot of cytoplasm—the male gamete only provides genetic information, and thus the cytoplasm from the original diploid cell can be equally divided between four sperm. Choice (C) says the opposite, so eliminate it. Oogenesis actually begins for female fetuses while they are still in utero, whereas for males spermatogenesis begins around the time of puberty (A). Women go through menopause which marks the halt of oogenesis, but in males, once spermatogenesis begins it continues throughout their lifetime.

5. **B** Trisomy 21, or Down Syndrome, is a genetic disorder most often due to non-disjunction of chromosome 21 during oogenesis. According to the graph, the risk of this disorder increases with a mother's age, so it is reasonable to assume that the risk of non-disjunction increases with a mother's age (B). Choice (A) says the opposite. Base mutations do not play a role in the etiology of Trisomy 21, so eliminate (C) and (D).

6. **D** Codominance means that both alleles are equally expressed, so both proteins would be expressed in the case of an individual with both the A and B blood type alleles. Choices (A) and (B) describe classical (Mendelian) dominance, which codominance is an exception to. There is no type of dominance that would explain "cancelling each other out," so eliminate choice (C).

7. **A** In general, the human endocrine system releases hormones into the bloodstream, and exocrine function refers to the release of enzymes from a gland into a duct. Eliminate (B), (C), and (D) as they do not reflect this relationship.

8. **A** The curve for exercise is right-shifted, and for any given oxygen concentration the saturation of hemoglobin by oxygen percentage is lower than that for the normal curve. This means that less oxygen is bound to hemoglobin and that more is released, or delivered, to muscle cells. Eliminate choice (B) because the saturation of hemoglobin by oxygen increases according to the graph. The conditions listed indicate that the concentration of carbon dioxide increases, so eliminate choice (D). And as stated above, oxygen delivery to muscles increases in the right-shifted curve, so eliminate choice (C), too.

9. **D** New red blood cells are created in the bone marrow, and the immune system does not play a role in this activity. Choices (A), (B), and (C) are all different activities that the immune system does perform.

10. **A** Filtrate is pushed from the glomerulus into Bowman's capsule and then enters the proximal convoluted tubule; from there, it proceeds through the loop of Henle, distal convoluted tubule, and collecting duct. By the end of the collecting duct, this fluid is called urine, and drains into the ureters, down to the bladder, and then finally leaves the body through the urethra.

11. **B** Reabsorption refers to the movement of molecules and ions from the filtrate back into the bloodstream, whereas secretion refers to movement from the bloodstream into the filtrate along the course of the tubules of the nephron. Thus, I and II are true; eliminate choices (A) and (C). While alterations in reabsorption and secretion can play a role in helping to correct blood pH if it is abnormal, these processes both constantly occur even when pH is normal in order to help maintain normal pH as well as regulate water and salt balance in the body. III is false, so eliminate choice (D).

12. **A** Sweat glands secrete water and ions onto the skin's surface and sweat glands are found in the dermis, below the epidermis and the layer of dead skin cells known as the stratum corneum; eliminate (B) and (C). Subcutaneous tissue is found below the dermis (D).

13. **D** The function of the myelin sheath surrounding an axon is to propagate the action potential down the axon by allowing it to jump from one spot to another down the axon without needing to depolarize every single spot on the axon's cell membrane. By insulating the axon in this way, the impulse can travel faster (A). The uncovered spots, called nodes of Ranvier, are the places where the impulse jumps, and this jumping from node to node is known as saltatory conduction, eliminating (B) and (C).

14. **D** Acetylcholinesterase is the enzyme that breaks down acetylcholine, the neurotransmitter used in the synapses of the parasympathetic nervous system; the suffix "-ase" means that this is an enzyme. Neurotransmitters are released from the end of an axon when calcium enters its terminal end and is picked up quickly by the dendrites on the post-synaptic neuron, eliminating (A) and (B). Neurotransmitters released at a neuromuscular junction can stimulate muscle contraction (C).

15. **C** Skeletal muscle movement is a voluntary activity, which is controlled by the somatic nervous system. The autonomic nervous system controls involuntary activities and is further subdivided into the parasympathetic and sympathetic nervous systems.

16. **B** The period following an action potential is known as the refractory period. Because the cell membrane becomes hyperpolarized at the end of the action potential, it is more negative than normal resting potential, and thus it is more difficult than usual to reach the threshold for depolarization. The cell membrane potential is positive following depolarization, not at the end of the action potential (A). While sodium channels are inactive at some point during the action potential, by the end of the action potential they are reset and are able to open (C). Potassium channels do not inactivate (D).

17. **C** After an egg is released from the ovary, it travels down the fallopian tube into the uterus. If it is fertilized by a sperm, this most often occurs in the fallopian tube. The zygote then travels down the rest of the fallopian tube into the uterus, where it implants. While fertilization can occur in the uterus, this does not occur as often as fertilization in the fallopian tube and would not allow for ideal timing of zygote early development prior to implantation (A). Fertilization must occur prior to implantation, so choice (B) does not make any sense. Fertilization cannot occur prior to the egg being released from the ovary (D).

18. **C** According to the poster, the embryonic period ends after 8 weeks, and then the dots indicating the most common site of birth defects are found only in the fetus's brain. The ears, eyes, and heart are all common sites of birth defects at different weeks during embryonic development.

19. **B** The taiga biome is found in northern forests and is very cold; thus, its plant life consists of wind-blown conifers that are stunted in growth, and its animal life includes caribou, wolves, moose, bear, rabbits, and lynx. Grasslands have hot summers and cold winters, with grasses as the main form of plant life (A). The temperate deciduous forest contains deciduous trees whose leaves fall off in the winter (C). The tundra has very few trees and instead has mostly grasses and flowers (D).

20. **A** Hurricanes are natural phenomena that will affect a population regardless of that population's density and are therefore density-independent. Resource depletion, competition, and predation all affect a population and the extent of this effect depends on the population's density, and therefore these are all density-dependent factors.

21. **D** Energy flow, biomass, and numbers of members within an ecosystem track together within an ecological pyramid, so those populations with the largest number of members belong at the bottom of the period, moving upward to the population with the least number of members. Therefore, the ecological pyramid for this ecosystem would look like this.

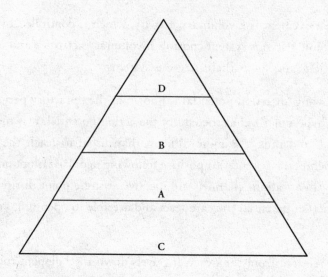

If population B suddenly increased in numbers, these increased numbers would result in an increase in the consumption of population B. Since this population represents secondary consumers, they would consume from population A, and the overall biomass of population A would be expected to decrease. The amount of energy and the biomass of population D might increase, at least at first, as this population consumes from population B and thus would have more resources. The amount of energy in population A would decrease along with its biomass.

22. **C** The primary role of B-cell mediated (humoral) immunity is to generate antibodies in response to foreign antigens. They do not release chemicals to destroy infected cells, that is the role of T cells.

23. **C** The glomerulus filters out small molecules such as urea, water, and salts. However, protein and cells should never be present in the filtrate of the kidneys. This is why urine tests are performed to diagnose kidney infections and kidney failure.

24. **D** Antidiuretic hormone (ADH) results in increased water reabsorption and would therefore lower the osmolarity (solute concentration) of the blood. An inhibitor of ADH, such as alcohol, would result in the opposite effect (decreased water retention and increased blood osmolarity).

25. **B** Following filtration in the glomerulus, further filtration and concentration of the filtrate first occurs in the proximal (in reference to the glomerulus) convoluted tubule, then the loop of Henle, followed by the distal convoluted tubule, and lasting collecting duct.

Chapter 11
Big Idea 1 Drill 2

BIG IDEA 1 DRILL 2

Multiple-Choice Questions

1. Which of the following would increase genetic variation in a population?

   (A) Inbreeding
   (B) Small population size
   (C) Genetic drift
   (D) Nonrandom mating

2. An adaptation is a genetic variation that ultimately results in

   (A) longer life span for an organism
   (B) competitive advantage in finding food
   (C) increased ability to store food
   (D) greater reproductive success

3. Which of the following does NOT provide evidence for the relatedness of all organisms?

   (A) DNA and RNA
   (B) Cytoskeleton
   (C) Genetic code
   (D) Metabolic pathways

4. Scientific evidence supports which of the following regarding the origin of life on Earth?

   (A) It was possible to form organic molecules from inorganic molecules prior to the presence of life on Earth.
   (B) The earliest life on Earth dates to about 4.6 billion years ago.
   (C) Each current domain (Archaea, Bacteria, and Eukarya) has a unique ancestral origin of life.
   (D) The original environment of the Earth was hospitable to life.

5. Stanley Miller conducted many experiments attempting to model Earth's early atmosphere in the 1950s. He found that by combining ammonia, hydrogen, methane, and water, and adding electricity, he was able to synthesize some amino acids. His results do NOT support which of the following hypotheses?

(A) Oxygen was a necessary element to allow for the creation of organic molecules on Earth.
(B) Early conditions on Earth could have allowed for the creation of molecules necessary to life.
(C) Early conditions on Earth were sufficient for the creation of organic molecules.
(D) Some sort of energy was necessary to allow for the creation of organic molecules on Earth.

6. Given the phylogenetic tree for six different species of bacteria shown below, which of the two species would you expect to have the greatest number of nucleotide differences in their rRNA subunits?

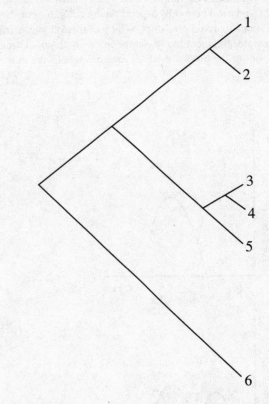

(A) 1 and 2
(B) 3 and 4
(C) 3 and 5
(D) 3 and 6

7. In a hypothetical city, the mayor decides to paint all of the buildings grey in order to make the city skyline look more uniform. Prior to this event, there was a wide range of light to dark colored buildings in the city, which was matched by a population of moths that ranged in coloration from light to dark. The moths depend on the coloration of the buildings to avoid predators. Which of the following graphs illustrates the most likely change to the moth population as a result of the human-made environmental change in the city?

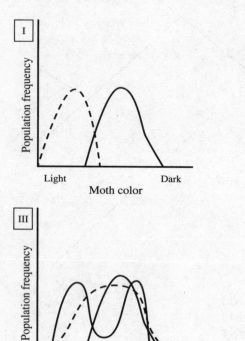

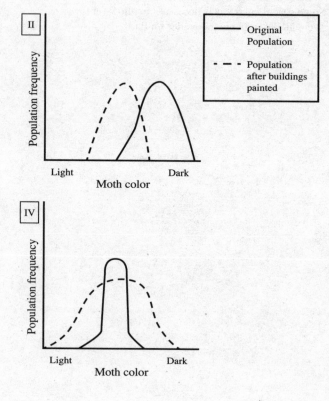

(A) Figure I
(B) Figure II
(C) Figure III
(D) Figure IV

8. Experimental evidence demonstrates that both eukaryotes and prokaryotes can utilize anaerobic metabolism and that the process is very similar in the two different types of organisms. This evidence best supports which of the following hypotheses?

(A) Anaerobic metabolism is a common energy-releasing process for prokaryotes and eukaryotes, suggesting that all prokaryotes and all eukaryotes thrive in anaerobic conditions.

(B) Anaerobic metabolism is the primary energy-releasing process for all prokaryotes and all eukaryotes, suggesting that both prokaryotes and eukaryotes carry out glycolysis in the cytosol.

(C) Anaerobic metabolism is a common energy-releasing process for prokaryotes and eukaryotes, suggesting that prokaryotes and eukaryotes share a common ancestor.

(D) Anaerobic metabolism is the primary energy-releasing process for all prokaryotes and all eukaryotes, suggesting that prokaryotes and eukaryotes underwent convergent evolution.

9. The electron transport chain in eukaryotes and prokaryotes

(A) occurs in the same place (within the mitochondrial matrix)

(B) occurs in the same place (across the inner mitochondrial membrane)

(C) occurs within the mitochondrial matrix in eukaryotes and in the cytoplasm in prokaryotes

(D) occurs across the inner mitochondrial membrane in eukaryotes and across the plasma membrane in prokaryotes

10. A feature of triglycerides not found in proteins is the presence of

(A) peptide bonds
(B) ester linkages
(C) nitrogen atoms
(D) oxygen atoms

11. Which is NOT a characteristic of eukaryotic cells?

(A) Circular double-stranded DNA
(B) Membrane-bound mitochondria
(C) Plasma membrane made up of lipids and proteins
(D) Two ribosomal subunits that come together to synthesize polypeptides

12. What best describes the role of the population unit in the process of evolution?

(A) A population is the level at which natural selection occurs.

(B) A population is the only level at which genetic changes can occur.

(C) Populations are the only groups in which a gene pool always remains constant over time.

(D) Populations are the groups in which evolution occurs over time.

13. The potential energy carried by reduced electron carriers during aerobic cellular respiration is ultimately used to

(A) make ATP
(B) break down glucose
(C) oxidize glucose
(D) break down fats

14. All of the following structures are homologous to a bird wing EXCEPT

(A) a bat wing
(B) a human arm
(C) an insect wing
(D) a whale flipper

15. Which of the following pieces of information would best help to support the theory that mitochondria were originally derived from simple prokaryotic cells?

(A) Mitochondria are membrane-bound organelles.

(B) Mitochondria have a single circular double-stranded DNA genome.

(C) Mitochondria have multiple linear double-stranded DNA chromosomes.

(D) Mitochondria use the same enzymes for transcription and translation as everywhere else in a eukaryotic cell.

16. Viruses that infect animal cells escape the host cell by a process called "budding," whereby the virus buds off from the host cell and is enveloped by a piece of the host's plasma membrane. Viruses that infect bacterial cells are not able to use budding and instead must lyse the bacterial cell in order to escape. What best explains this difference?

(A) Viruses that infect animal cells have more available ATP.

(B) Viruses that infect bacterial cells have more available ATP.

(C) Animal cells have more hospitable plasma membranes.

(D) Bacterial cells have plasma membranes surrounded by a cell wall.

17. Which of the following examples does NOT provide evidence for evolution?

(A) Similar traits in animals living in geographically distinct areas

(B) Gill slits in both human embryos and whale embryos

(C) Overlapping genetic codes for all species

(D) Fossils of horses that look the same as contemporary horses

18. In a hypothetical mouse population, long tails are dominant over short tails in a classically dominant fashion. If 75% of mice have long tails, what is the frequency of the allele for short tails?

(A) 5%

(B) 15%

(C) 25%

(D) 75%

19. Which of the following is true of diploid cells?

(A) The four daughter cells resulting from meiosis are diploid cells.

(B) In humans, diploid cells have 42 chromosomes.

(C) Human gametes are an example of diploid cells.

(D) The two daughter cells resulting from mitosis are diploid cells.

20. Which part(s) of meiosis are responsible for helping to increase genetic diversity in a population?

    I. Random arrangement of tetrads during metaphase I
    II. Crossing over
    III. Separation of sister chromatids during anaphase II

(A) I only
(B) II only
(C) I and II
(D) I, II, and III

21. According to the simplified "Tree of Life" (based on genome sequencing) shown below, which of the two types of organisms are most closely related evolutionarily?

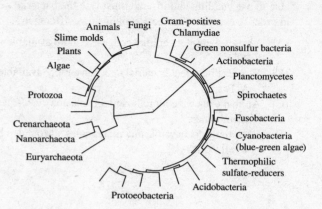

(A) Actinobacteria and Acidobacteria
(B) Protozoa and Algae
(C) Animals and Euryarchaeota
(D) Cyanobacteria and Thermophilic sulfate-reducers

Chapter 12
Big Idea 1
Drill 2 Answers and Explanations

ANSWER KEY

1. C
2. D
3. B
4. A
5. A
6. D
7. D
8. C
9. D
10. B
11. A
12. D
13. A
14. C
15. B
16. D
17. D
18. C
19. D
20. C
21. D

ANSWERS AND EXPLANATIONS

Multiple-Choice Questions

1. **C** Genetic drift refers to the change in frequency of alleles in a population due to randomness and will lead to increasing genetic variation due to chance. Inbreeding, small population size, and non-random mating will all cause a decrease in genetic variation.

2. **D** Evolutionary fitness is ultimately measured by an organism's reproductive success. While longer life span, competitive advantage in finding food, and greater ability to store food are all characteristics that could lead to greater reproductive success, this is not necessarily true; only reproductive success will define an adaptation, which is a genetic variation favored by selection and conferring an advantage to an organism.

3. **B** While the cytoskeleton provides structural evidence for the relatedness of all eukaryotes, prokaryotes do not have a cytoskeleton. DNA and RNA carry genetic information for all organisms through the processes of transcription, translation, and replication (A). All modern organisms share the same universal genetic code, where a sequence of three amino acids (called a codon) code for a specific amino acid (C). Many metabolic pathways, such as glycolysis, are conserved across all domains (Archaea, Bacteria, and Eukarya) (D).

4. **A** Chemical experiments have demonstrated that it is in fact possible for inorganic molecules to form organic molecules with the addition of energy. While the Earth was formed about 4.6 billion years ago, the first evidence of life in the form of fossils are 3.5 billion years old (B). Molecular and genetic evidence point to a common ancestral origin of life for all current domains (C). The original environment of the Earth was inhospitable to life, and it was only over time and changes in the atmosphere that life could exist on Earth (D).

5. **A** Because Miller was able to synthesize some amino acids, which are organic molecules, without oxygen, his experiment does not support the hypothesis that oxygen was necessary for the creation of organic molecules. Because amino acids are necessary to life, and Miller was attempting to replicate early conditions on Earth, choices (B) and (C) are supported by his experiment. In his experiment, Miller added electricity, supporting the hypothesis that energy was needed to create organic molecules (D).

6. **D** Phylogenetic trees are graphical models of evolutionary history that demonstrate the evolutionary relatedness of different species. More closely related species would have fewer differences in their rRNA subunits. The species of bacteria least closely related to the others as demonstrated in the figure is species 6; species 1 and 2, 3 and 4, and 3 and 5 are all more closely related according to the phylogenetic tree than species 3 and 6.

7. **D** Figure IV represents the most likely change in the moth population due to painting the buildings grey. Because the buildings used to range in color from light to dark, the original moth population would also be expected to range in color from light to dark (and the question stem also indicates this); eliminate choices (A) and (B). When all of the buildings are painted grey, you can expect that the moths at the extremes of coloration (light and dark) would be eaten by predators more often than the medium-colored moths, so eliminate choice (C).

8. **C** Because both prokaryotes and eukaryotes can undergo anaerobic metabolism this is a common energy-releasing process for both types of organisms, but this does not mean that it is the primary energy-releasing process for all prokaryotes and all eukaryotes—eliminate choices (B) and (D). Having anaerobic metabolism in common does not indicate that all prokaryotes and eukaryotes thrive without oxygen, however, so eliminate choice (A). This common process does suggest that these two types of organisms likely share a common ancestor.

9. **D** The protein carriers that pass electrons along the electron transport chain are found in the inner mitochondrial membrane of eukaryotes and in the plasma membrane of prokaryotes. Prokaryotes do not have mitochondria, so eliminate choices (A) and (B). The electron transport chain is responsible for creating a proton gradient, so it must occur across a membrane separating the area of high proton concentration from the area of low proton concentration—eliminate choice (C).

10. **B** A triglyceride is made up for three fatty acids bound to a glycerol molecule via an ester linkage. Peptide bonds link amino acids together to make a protein (A). Nitrogen atoms and oxygen atoms are present in both triglycerides and proteins, (C) and (D).

11. **A** Eukaryotic cells contain linear double-stranded DNA molecules known as chromosomes; prokaryotic cells contain a single circular double-stranded DNA molecule. Membrane bound mitochondria, a plasma membrane of lipids and proteins, and two ribosomal subunits needed for translation are all important eukaryotic cellular components.

12. **D** The population unit is described as the unit of evolution because changes in the genetic makeup of populations can be measured over time. Natural selection affects individual organisms, not populations (A). Likewise, genetic changes occur at the individual level (B). A gene pool may remain constant within a population under Hardy-Weinberg equilibrium, but this is not always the case (C).

13. **A** The reduced electron carriers (NADH and $FADH_2$) made during cellular respiration are carrying potential energy in the form of electrons, which are used to establish a proton gradient that ultimately is used by the enzyme ATP synthase to make ATP. The breakdown of glucose and fats are energy-releasing processes that contribute to making ATP. Glucose is oxidized during cellular respiration, but is not dependent on reduced electron carriers—rather, it is the oxidation of glucose that allows for the reduction of electron carriers.

14. **C** While an insect wing and a bird wing are analogous structures, they are phylogenetically indepen-dent and thus not homologous structures. Bat wings, human arms, and whale flippers all evolved from some structure in a common ancestor.

15. **B** The endosymbiotic theory proposes that mitochondria were derived from prokaryotic cells, and one piece of evidence in support of this theory is the circular DNA genome of mitochondria, which is very similar to the single circular double-stranded DNA genome of bacteria. If mitochondria had multiple linear chromosomes, this would not support a bacterial origin (C). All eukaryotic organelles are membrane-bound, so this fact does not provide evidence for a bacterial ancestor of mitochondria (A). If mitochondria used the same enzymes for transcription and translation as the rest of the eukaryotic cells, this would not support a separate origin for mitochondria (D).

16. **D** Because bacterial cells are surrounded by a cell wall, the viruses that infect them are not able to escape from the host cell by budding—instead, they must lyse the cell. Variations in the amount of available ATP would not make a difference in exit strategy, as a cell wall will prevent a barrier to budding no matter what the energy resources of the virus, eliminating (A) and (B). It is not the hospitability of the animal cell membranes, but the fact that they do not have a cell wall, that allows budding for viruses infecting animal cells (C).

17. **D** Fossils of horses that show that they were the same then as horses today would not provide evidence for evolution, as it would seems horses has not changed or evolved in the interim time. Animals living in different areas sharing similar traits, ubiquitous gill slits in many different species' embryos, and overlaps in the genetic code for different species all provide evidence for evolution.

18. **C** Using the Hardy-Weinberg equation for genotype frequency ($p^2 + 2pq + q^2 = 1$), you can plug in 0.75 for $p^2 + 2pq$ (because that expression represents the dominant phenotypes—homozygous dominant and heterozygous genotypes). That will give the equation $q^2 = 0.25$, which simplifies to $q = 0.05$. Because q represents the recessive allele frequency, the frequency of the allele for short tails is 5%.

19. **D** Mitosis produces two identical daughter cells, which are both also identical to the original parent cell. Diploid cells undergo mitosis, and thus the two daughter cells are also diploid. The four cells resulting from meiosis are haploid (A). In humans, diploid cells have 46 chromosomes (B). Human gametes are an example of haploid cells (C).

20. **C** Both the process of crossing over and the random arrangement of tetrads in metaphase I promote genetic diversity in a population by allowing different combinations of genes and chromosomes in offspring than in their parents. Sister chromatids are exactly alike, so their separation does not promote any genetic diversity (III).

21. **D** This "Tree of Life" is similar to a phylogenetic tree, and cyanobacteria and thermophilic sulfate-reducers are closest evolutionarily as they come from a common branch off of the tree. All of the other proposed pairings are further apart as shown on the tree.

Chapter 13
Big Idea 2 Drill 2

BIG IDEA 2 DRILL 2

Multiple-Choice Questions

1. Normal cellular death that is necessary for normal development and morphogenesis is referred to as

   (A) apoptosis
   (B) cancer
   (C) differentiation
   (D) maturation

2. Which of the following could be affected by a lack of energy in the environment?

   I. An individual organism
   II. A population
   III. An ecosystem

   (A) I only
   (B) I and II
   (C) I and III
   (D) I, II, and III

3. Which process would be least likely to occur in the absence of inorganic chemicals?

   (A) Cellular respiration
   (B) Chemosynthesis
   (C) Fermentation
   (D) Photosynthesis

4. Which of the following molecules is LEAST likely to require a transport protein to move across a cell membrane?

   (A) Water
   (B) Hydrogen ions
   (C) Carbon dioxide
   (D) Chloride ions

5. When a pregnant woman is ready to give birth, the fetus pushes on her uterus, which causes a hormone called oxytocin to be released from the pituitary gland. Oxytocin causes contraction of the uterus, which results in the fetus pushing even more on her uterus, and the release of more oxytocin from the pituitary gland. This series of events is an example of

(A) enzymatic regulation
(B) feed-forward control
(C) negative feedback
(D) positive feedback

6. Which of the following correctly identifies and describes the waste products of cellular respiration?

(A) Autotrophs produce water and carbon dioxide as waste products of cellular respiration; these organisms reuse the water and give off carbon dioxide, which can be utilized by heterotrophs to produce carbohydrates.

(B) Heterotrophs produce water and carbon dioxide as waste products of cellular respiration; these organisms reuse the water and give off carbon dioxide, which can be utilized by autotrophs to produce carbohydrates.

(C) Autotrophs produce oxygen and glucose as waste products of cellular respiration; these organisms reuse the glucose and give off oxygen, which can be utilized by heterotrophs to produce carbohydrates.

(D) Heterotrophs produce oxygen and glucose as waste products of cellular respiration; these organisms reuse the glucose and give off oxygen, which can be utilized by autotrophs to produce carbohydrates.

7. Which of the following elements is/are common to both nucleic acids AND proteins?

   I. Carbon
   II. Nitrogen
   III. Phosphorous

(A) I only
(B) I and II
(C) II and III
(D) I, II, and III

8. Most cellular activities are unfavorable reactions with a negative change in free energy. How are these endergonic reactions able to occur?

(A) Free energy lost to entropy exceeds energy input in a cell.
(B) Cellular processes that increase entropy are coupled to endergonic reactions in a cell.
(C) Cells are special systems that are able to violate the second law of thermodynamics.
(D) Exergonic reactions are coupled to endergonic reactions in a cell.

9. What is the main strategy employed by endotherms to maintain homeostatic body temperature?

  (A) Utilization of external thermal energy
  (B) Utilization of thermal energy incidentally produced as a byproduct of metabolism
  (C) Utilization of muscular exertion mechanisms such as shivering
  (D) Utilization of special metabolic processes such as breakdown of brown fat

10. Which of the following is most likely to have the highest metabolic rate per unit body mass?

  (A) An earthworm
  (B) An eagle
  (C) A human
  (D) An elephant

11. If an organism is able to gather more free energy than it requires to maintain normal functions, which of the following may result?

  I. Energy storage
  II. Growth
  III. Build-up of toxic ATP levels

  (A) I only
  (B) I and II
  (C) I and III
  (D) II and III

12. Which of the following processes do NOT require oxygen?

  I. Photosynthesis
  II. Chemosynthesis
  III. Fermentation

  (A) III only
  (B) I and II
  (C) I and III
  (D) I, II, and III

13. Oxygen plays a role in cellular respiration most similar to the role of

(A) NADP⁺ in photosynthesis
(B) carbon dioxide in photosynthesis
(C) NADP⁺ in fermentation
(D) glucose in fermentation

14. The inner mitochondrial membrane establishes a gradient of protons as electrons are transported to different proteins during one of the final steps of cellular respiration. During photosynthesis in plant cells, a similar gradient is established

(A) by the thylakoid membrane via the Calvin cycle
(B) by the thylakoid membrane via the electron transport chain
(C) by ATP synthase via the electron transport chain
(D) by ATP synthase via the Calvin cycle

15. A pyruvate molecule produced by glycolysis

(A) has the same number of carbon atoms as glucose
(B) is transported from the cytoplasm to the mitochondrial matrix
(C) is transported from the mitochondrial matrix to the intermembranous space of the mitochondrion
(D) can enter the Krebs cycle

16. Which of the following establishes concentration gradients that can then be used to help with dynamic homeostasis in cells?

(A) Active transport
(B) Diffusion
(C) Facilitated diffusion
(D) Osmosis

17. Which of the following components of the cell membrane contribute the most to its role in separating the intracellular from extracellular environments?

(A) Cholesterol
(B) Glycoproteins
(C) Phospholipids
(D) Transport proteins

18. A red blood cell is approximately 1% sodium by concentration. How would you describe a red blood cell in a beaker of pure water?

(A) The red blood cell is hypotonic to the surrounding water.
(B) The red blood cell is hypertonic to the surrounding water.
(C) The red blood cells is isotonic to the surrounding water.
(D) The red blood cell does not have any tonicity compared to the surrounding water.

19. Some scientists have argued that any close interaction between two organisms can be truly neutral for either of the organisms, and that both must either benefit, or one benefits while the other harmed, even if the benefit or harm is very subtle and perhaps currently unknown. These scientists are arguing against the concept of

(A) commensalism
(B) mutualism
(C) parasitism
(D) predator-prey relationships

20. Which of the following is NOT true of mammalian immune systems?

(A) Mammalian immune systems include nonspecific components.
(B) Mammalian immune systems include specific components.
(C) Mammalian immune systems include antibodies that are specific to particular antigens.
(D) Mammalian immune systems include defense mechanisms against pathogens that are all heritable from parent to offspring.

21. Many flowering plants determine the proper time to flower by detecting seasonal changes in the length of day or night. This is an example of

(A) gravitropism
(B) photoperiodism
(C) phototropism
(D) thigmotropism

Questions 22-24 refer to the following.

Carbonic anhydrase is a catalytic enzyme, which converts carbon dioxide and water into bicarbonate ions and protons (as shown below). The reaction is critical for maintenance of blood pH and for increasing the solubility of carbon dioxide in the blood.

22. An athlete is purposefully hyperventilating before beginning a deep dive to increase his ability to hold his breath. Which of the following changes would be expected to occur to his blood pH?

(A) His blood pH would increase due to an increase in carbon dioxide in the blood and thus more H^+ is present by the formation of bicarbonate.
(B) His blood pH would increase due to a decrease in carbon dioxide in the blood and thus less H^+ is present by the formation of bicarbonate.
(C) His blood pH would decrease due to an increase in carbon dioxide in the blood and thus more H^+ is present by the formation of bicarbonate.
(D) His blood pH would decrease due to a decrease in carbon dioxide in the blood and thus less H^+ is present by the formation of bicarbonate.

23. The pH of the blood is most nearly which of the following?

(A) 6.4
(B) 6.8
(C) 7.4
(D) 7.8

24. Which of the following best explains the properties of the blood immediately prior to reaching the lungs?

(A) The blood is oxygenated and carbon dioxide rich.
(B) The blood is deoxygenated and carbon dioxide rich.
(C) The blood is oxygenated and carbon dioxide poor.
(D) The blood is deoxygenated and carbon dioxide poor.

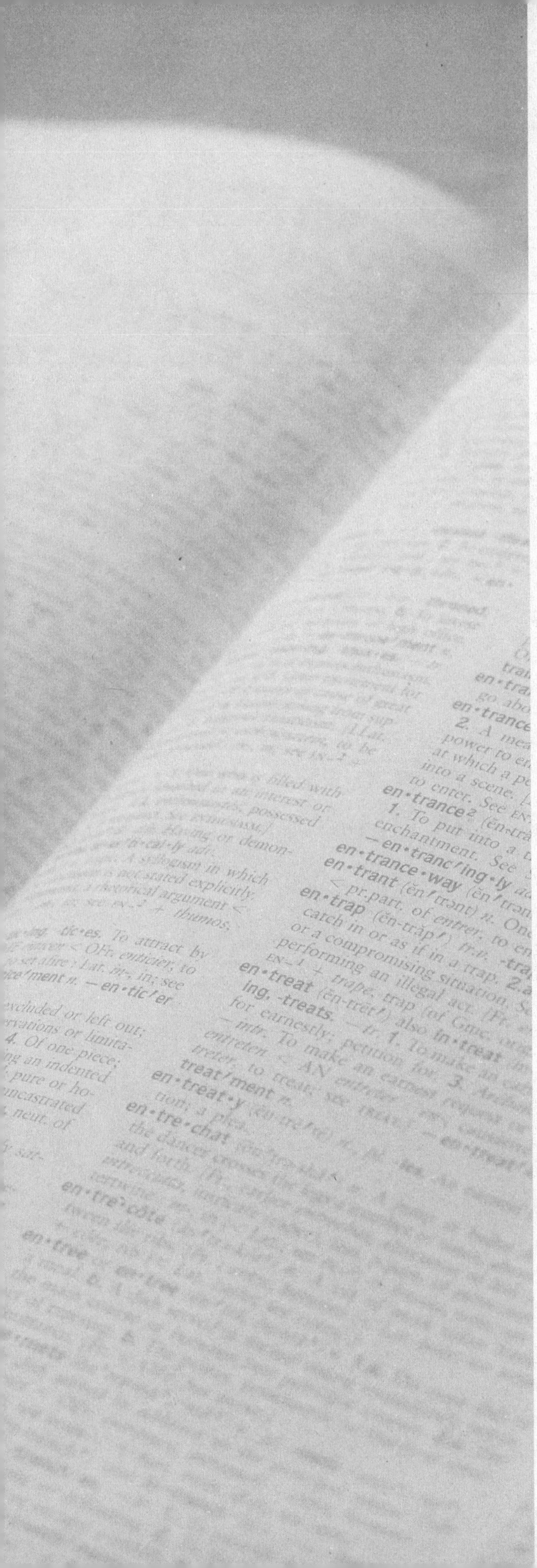

Chapter 14
Big Idea 2
Drill 2 Answers and Explanations

ANSWER KEY

1. A
2. D
3. B
4. C
5. D
6. B
7. B
8. D
9. B
10. A
11. B
12. D
13. A
14. B
15. B
16. A
17. C
18. B
19. A
20. D
21. B
22. B
23. C
24. B

ANSWERS AND EXPLANATIONS

Multiple-Choice Questions

1. **A** Apoptosis is the term for programmed cell death, which is an essential part of normal development and morphogenesis in organisms. Cancer is an abnormal process where cells typically lose the ability to undergo apoptosis (B). Differentiation and maturation are also important processes to normal development and morphogenesis but refer to the specialization of cells rather than their death, eliminating (C) and (D).

2. **D** An energy deficiency would not only be problematic at the level of individual organisms but also could hurt a population of organisms as well as the entire ecosystem.

3. **B** Chemosynthesis is used by autotrophs to capture free energy from inorganic chemicals. Cellular respiration and fermentation are processes that utilize sugars (organic molecules) to harvest free energy to produce ATP, eliminating (A) and (C). Photosynthesis utilizes the free energy from sunlight to produce carbohydrates (D).

4. **C** Carbon dioxide is a small hydrophobic molecule, so it is able to easily traverse the hydrophobic lipid tails that make up the majority of the plasma membrane. Water, hydrogen ions, and chloride ions all have polarity and thus are hydrophilic—these molecules all require a transport protein to move between the intracellular and extracellular environments.

5. **D** This series of events is a good example of positive feedback, whereby a response is amplified—in this case, uterine contraction. This specific positive feedback loop is stopped only by the delivery of the baby. While enzymatic regulation is often dependent on feedback, no enzymes are mentioned in this situation (A). Feedforward control refers to a certain molecule in a pathway "turning on" an enzyme necessary utilized later on in the pathway (B). Negative feedback loops allow for optimum levels of products by "turning off" enzymes once a certain desired level of product has been produced (C).

6. **B** Heterotrophs harvest free energy from organic molecules produced by other organisms via the process of cellular respiration, while autotrophs capture free energy from the environment through chemosynthesis and photosynthesis—eliminate (A) and (C). The waste products of cellular respiration are carbon dioxide and water, while glucose and oxygen are necessary for cellular respiration; eliminate choice (D).

7. **B** Carbon and nitrogen are found in nucleic acids and proteins. Phosphorous is found in nucleic acids (and phospholipids and ATP) but not proteins.

8. **D** Exergonic reactions that are energetically favorable are used to drive endergonic reactions that are energetically unfavorable in a cell; the most common exergonic reaction used is ATP → ADP. Energy input into a cell must exceed the free energy lost to entropy in a cell in order for a cell to remain alive (A). Endergonic reactions must be coupled to cellular processes that decrease entropy (and therefore result in a positive change in free energy) (B). Cells do not violate but obey the second law of thermodynamics—entropy increases over time (C).

9. **B** Endotherms, or "warm-blooded" animals, rely on the thermal energy produced from routine metabolism to regulate and maintain body temperature. In contrast, ectoderms, or "cold-blooded" animals, rely almost purely on ambient heat (A). Some endotherms employ special methods to create heat under extremely cold conditions, such as shivering or metabolizing brown fat, but these are not their main methods to maintain body temperature, eliminating (C) and (D).

10. **A** In general, the smaller a multicellular organism, the higher the metabolic rate per unit body mass. Out of all of the examples, an earthworm has the lowest body mass.

11. **B** An excess of acquired free energy compared with required free energy expenditure will result in energy storage or growth of the organism. There is no mechanism by which ATP build-up would cause toxicity; rather, organisms will store or utilize excess energy.

12. **D** Photosynthesis, chemosynthesis, and fermentation all may occur in the absence of oxygen. Chemosynthesis refers to the capture of free energy from small inorganic molecules in the environment.

13. **A** Oxygen acts as the final electron acceptor in cellular respiration, as does $NADP^+$ in photosynthesis. Carbon dioxide is one of the reactants in photosynthesis, but is not an electron acceptor; glucose is a reactant in fermentation, but also is not an electron acceptor, so eliminate (B) and (D). $NADP^+$ does not play a role in fermentation (C).

14. **B** The thylakoid membrane is the internal membrane of chloroplasts, and it is along this membrane that the electron transport chain occurs and across this membrane that the proton gradient is established in photosynthesis. The Calvin cycle utilizes the energy captured in the light reactions as ATP and NADPH to produces carbohydrates (A). ATP synthase does not establish the proton gradient but utilizes the proton gradient to synthesize ATP, (C) and (D).

15. **B** Pyruvate is made in the cytoplasm and is then transported to the mitochondrial matrix in order to continue through the process of cellular respiration. Pyruvate does not enter the intermembranous space (C). A pyruvate molecule has three carbon atoms, which is half the number as a glucose molecule (A). Pyruvate cannot directly enter the Krebs cycle but has to first be converted to acetyl CoA (D).

16. **A** Active transport requires energy and transport proteins to move molecules against their concentration gradient across a membrane, thus establishing a useful gradient that can be coupled to other processes necessary for dynamic homeostasis. Diffusion, facilitated diffusion, and osmosis are all

examples of passive transport, which are all characterized by spontaneous movement of molecules down a concentration gradient—thus, no meaningful concentration gradient is established by these processes.

17. **C** Phospholipids are polar and give the cell membrane its hydrophilic and hydrophobic properties—they form the bilayer that separates the internal environment from the external environment of the cells. Cholesterol, glycoproteins, and transport proteins are all essential parts of the cell membrane but fulfill other specific roles and do not play as large of a role as a barrier to the external environment.

18. **B** Tonicity is a comparative measure that relates the number of particles, or amount of solute, from one thing to another. In this case, the red blood cell has more solute (sodium) than the surrounding solution (pure water), so it is hypertonic to the solution.

19. **A** Commensalism refers to a relationship between two organisms where one organism benefits without affecting the other, which the scientists described do not think can occur. Mutualism refers to a relationship where both organisms benefit, while parasitism occurs when one organism benefits while the other is harmed. Predator-prey relationships describe when one organism eats another organism.

20. **D** Mammals develop specific immunity to particular pathogens over the course of their lifespan, via cell-mediated and humoral immunity; offspring are not born with the same immunities that their parents have developed, but must develop this specific immunity on their own. Mammalian immune systems include nonspecific and specific components, and the ability to produce antibodies to antigens is a part of specific immunity.

21. **B** Photoperiodism refers to the response of organisms to the relative lengths of light and dark periods. Gravitropism refers to how plants respond to gravity (A). Phototropism refers to a plant's response to sunlight, such as bending toward light, but not to periods or lengths of time of sunlight (C). Thigmotropism refers to how plants respond to touch (D).

22. **B** During hyperventilation, the body is rid of carbon dioxide and gains a surplus of oxygen. Swimmers and particularly divers will often take many quick successive breaths to purposefully hyperventilate to rid the body of carbon dioxide to increase the length of time they may travel without surfacing. In ridding the body of carbon dioxide, the equilibrium shown will shift to the left (according to Le Chatelier's principle) resulting in a reduction in the concentration of the hydrogen ion (H^+). Because there will be less H^+ present, the blood pH will effectively increase (become more alkaline or basic).

23. **C** The pH of the blood is nearly 7.4, due in large part to the equilibrium (and buffer system) of carbon dioxide and bicarbonate in the blood.

24. **B** Prior to reaching the lungs, the blood contains a higher concentration of carbon dioxide (picked up as a waste product of cellular respiration in the tissues) and a lower concentration of oxygen (which had been used in the electron transport chain for cellular respiration in the tissue). The purpose of the lungs is to reverse the concentrations of these gases in the blood.

Chapter 15
Big Idea 3 Drill 2

BIG IDEA 3 DRILL 2

Multiple-Choice Questions

1. Mitosis is used by organisms in order to

   (A) produce identical progeny, allowing for gamete formation
   (B) produce identical progeny, allowing for growth, replacement of cells, or asexual reproduction
   (C) produce differential progeny, allowing for gamete formation
   (D) produce differential progeny, allowing for growth, replacement of cells, or asexual reproduction

2. Which of the following can make up the heritable information for viruses?

   I. Single-stranded DNA
   II. Double-stranded DNA
   III. Single-stranded RNA

   (A) II only
   (B) I and II
   (C) I and III
   (D) I, II, and III

3. Which of the following accurately describes the process of DNA replication?

   (A) Conservative
   (B) Dispersive
   (C) Non-conservative
   (D) Semi-conservative

4. Which of the following violates the typical rules for flow of genetic information in cells?

   (A) Anaerobic bacteria
   (B) Bacteriophages
   (C) Plant cells
   (D) Retroviruses

5. Which nitrogenous bases take up the most room within the DNA double-helix?

(A) Adenine and thymine
(B) Adenine and guanine
(C) Cytosine and thymine
(D) Thymine and uracil

6. Which of the following is NOT true of the process of translation?

(A) It is coupled with transcription in prokaryotic organisms.
(B) It is coupled with transcription in eukaryotic organisms.
(C) It requires a ribosome with rRNA to occur.
(D) It requires tRNA to bring amino acids to contribute to the growing protein.

7. Which of the following occur(s) during interphase?

I. Synthesis of DNA
II. Growth
III. Breakdown of the nuclear envelope

(A) I only
(B) I and II
(C) II and III
(D) I, II, and III

8. What would be the most likely outcome as a result of one set of homologous chromosomes not properly separating during meiosis?

(A) One of the resulting gametes would have three copies of that chromosome.
(B) One of the resulting gametes would have one copy of that chromosome.
(C) After fertilization, the resulting zygote could have three copies of that chromosome.
(D) After fertilization, the resulting zygote could have four copies of that chromosome.

9. In which of the following situations would Mendel's laws of segregation and independent assortment be LEAST upheld?

(A) Two genes on separate chromosomes
(B) Two genes very close to each other on the same chromosome
(C) Two genes on opposite ends of the same chromosome
(D) A gene on the X chromosome and a gene on the Y chromosome

10. A completely healthy man and woman have a son who is born with Disease A. This son marries a healthy woman, and none of their children have Disease A. However, one of their daughters (the original couple's grandchild) marries a healthy man, and their son (the original couple's great-grandchild) is born with Disease A. Which of the following best characterizes Disease A?

(A) Disease A is autosomal dominant
(B) Disease A is autosomal recessive
(C) Disease A is sex-linked dominant
(D) Disease A is sex-linked recessive

11. Which of the following processes allows for lateral exchange of genetic information in bacteria?

(A) Mitosis
(B) Meiosis
(C) Transduction
(D) Transfection

12. Viruses with RNA genomes have higher rates of mutation than viruses with DNA genomes. Which best explains this difference?

(A) RNA viruses go through the lytic cycle more quickly than DNA viruses.
(B) DNA viruses go through the lysogenic cycle more often than RNA viruses.
(C) RNA viruses lack replication error-checking mechanisms.
(D) DNA viruses lack replication error-checking mechanisms.

13. At some synapses, acetylcholine acts as a stimulatory neurotransmitter, causing an action potential in the post-synaptic (receiving) neuron. Acetylcholine esterase is an enzyme that breaks down acetylcholine and is found in the synaptic cleft (space between the two neurons). If acetylcholine esterase was NOT present in the synaptic cleft, what would most likely result?

    (A) More action potentials would be fired by the post-synaptic neuron.
    (B) Fewer action potentials would be fired by the post-synaptic neuron.
    (C) More action potentials would be fired by the pre-synaptic neuron.
    (D) Fewer action potentials would be fired by the pre-synaptic neuron.

14. Which of the following is/are always true of DNA replication?

    I. Synthesis of new DNA proceeds in a 5' to 3' direction.
    II. Synthesis of new DNA occurs in the nucleus.
    III. Each strand of the original DNA molecule is used as a template for the synthesis of a complimentary strand.

    (A) I only
    (B) I and II
    (C) I and III
    (D) I, II, and III

15. A mouse with black hair and a mouse with white hair mate and have a litter of ten baby mice. Which of the following observations would best support the hypothesis that hair color in mice is completely genetically controlled?

    (A) Each of the ten of the baby mice has a different shade of hair color.
    (B) Nine of the baby mice have white hair, and one has black hair.
    (C) Five of the baby mice have white hair, and five have black hair.
    (D) Nine of the baby mice have black hair, and one has white hair.

16. In an experiment, scientists used transformation by mixing together wild type *Acinetobacter* bacteria with a plasmid that carries a gene for antibiotic resistance to gentamicin (*gen*ʳ). They then plated this sample of *Acinetobacter* on agar and added gentamicin to one side of the plate. They also plated wild-type *Acinetobacter* on agar and added gentamicin to one half of that plate. The results are demonstrated below, with dots representing individual colonies of bacteria, and shaded areas representing such extensive growth of bacteria that individual colonies cannot be distinguished.

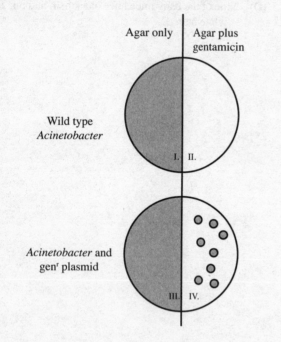

17. A human kidney serves to filter blood and excrete wastes as urine, while reabsorbing water and useful dissolved substances back into the blood stream. Antidiuretic hormone (ADH) is secreted in response to decreased blood pressure and/or blood volume and acts on the kidney to stimulate the insertion of more aquaporins into the collecting ducts, which allows for more water to be reabsorbed. If ADH secretion were blocked, which of the following would initially result?

(A) Less water would be consumed in order to compensate for the blockage of ADH secretion.

(B) More aquaporins would be inserted into the plasma membranes of collecting duct cells.

(C) More dilute urine would be produced.

(D) Negative feedback from ADH would increase.

18. Which of the following is the primary source of estrogen in human females?

(A) Anterior pituitary gland
(B) Corpus luteum of the ovary
(C) Endometrium of the uterus
(D) Maturing follicle of the ovary

Gentamicin-resistant *Acinetobacter* are growing on which of the following parts of the plates?

(A) I only
(B) I and II
(C) IV only
(D) III and IV

19. Which of the following is NOT true regarding mRNA processing?

(A) While eukaryotic mRNA must be processed, prokaryotic mRNA is not processed.

(B) mRNA processing occurs immediately following the molecules exit from the nucleus.

(C) mRNA processing requires an RNA-protein complex called a spliceosome.

(D) As a result of mRNA processing, introns are removed.

20. Which of the following accurately describes a similarity and a difference between the nervous system and the endocrine system?

(A) Both rely on hormones to send messages, but the nervous system works more rapidly than the endocrine system.

(B) Both systems work at the same rate, but the nervous system uses nerve impulses to send messages while the endocrine system utilizes hormones.

(C) Both systems work at the same rate, but the endocrine system utilizes temperature to send messages while the nervous systems uses nerve impulses.

(D) Both systems send messages to other systems to control body activities, but the endocrine system works more slowly than the nervous system.

21. Which of the following is NOT an example of cooperative behavior increasing population-level survival?

(A) A herd of wildebeest

(B) A school of fish

(C) Chimpanzees calling out in certain patterns when predators are present

(D) A rabbit running away from a predator

Chapter 16
Big Idea 3
Drill 2 Answers and Explanations

ANSWER KEY

1. B
2. D
3. D
4. D
5. B
6. B
7. B
8. C
9. B
10. D
11. C
12. C
13. A
14. C
15. C
16. D
17. C
18. D
19. B
20. D
21. D

ANSWERS AND EXPLANATIONS

Multiple-Choice Questions

1. **B** Mitosis results in fidelity in replication of DNA, resulting in two identical progeny cells; organisms grow, replace cells, and reproduce asexually via mitosis. Meiosis results in four differential progeny cells and is used for the production of gametes.

2. **D** Viruses are an exception in that their heritable information is not necessarily double stranded DNA; any combination of single- or double-stranded and RNA or DNA can make up viral genomes.

3. **D** DNA replication is semi-conservative, meaning that one strand of DNA serves as a template for a new daughter strand, and following replication there are two DNA double-stranded molecules, each made up of one parent strand and one daughter strand. If replication were conservative, one molecule would be made up of both parent strands, and the other molecule would be made up of both daughter strands (A). If replication were dispersive, pieces of both parent and daughter DNA would be in all of the strands of DNA following DNA replication (B). Non-conservative would imply that no parent DNA was in the new DNA molecules (C).

4. **D** Retroviruses break the rule of the traditional flow of genetic information (DNA → RNA → protein) by using an enzyme called reverse transcriptase, which synthesizes DNA from RNA. This is necessary because retroviruses have an RNA genome. All prokaryotic and eukaryotic cells have a DNA genome, and the flow of genetic information follows the normal route: DNA → RNA → protein. Bacteriophages are viruses that infect bacteria; these viruses typically have a double-stranded DNA genome, and none have reverse transcriptase (B).

5. **B** Purines (adenine and guanine) have a double-ring structure, while pyrimidines (cytosine, thymine, and uracil) have a single ring structure; thus, adenine and guanine take up more space within the DNA double-helix than pyrimidine bases (A, C). Uracil is only found in RNA, not DNA (D).

6. **B** In eukaryotes, transcription occurs in the nucleus while translation occurs in the cytoplasm, so these two processes are spatially and temporally distinct. Transcription and translation both occur in the cytoplasm of prokaryotic organisms, which allows for coupling of the two processes (A). Translation depends on ribosomes and tRNA in order to read mRNA and build a protein, eliminating (C) and (D).

7. **B** Cell growth and synthesis of DNA both occur during interphase. Breakdown of the nuclear envelope is one of the first events of mitosis, during prophase.

8. **C**   If one set of homologous chromosomes did not properly separate during meiosis, both chromosomes would end up in some gametes, while no copies of that chromosome would be in other gametes, (A) and (B). Following fertilization, one more copy of that chromosome would be added to the gametes, so the resulting zygote could have either three or one copy of that chromosome (D).

9. **B**   Genes that are adjacent and close to each other on the same chromosome typically stay together as a single unit, as it is unlikely that a crossing over event will occur between them. The further away two genes are on the chromosome, the more likely there will be a crossing over event that allows them to segregate and assort independently (C). Genes on completely different chromosomes always segregate and assort independently, (A) and (D).

10. **D**   Because only males are afflicted with Disease A and it skips generations, it is sex-linked recessive. Specifically, Disease A is X-linked recessive, which explains how women can carry the trait on one of their X chromosomes and pass it onto their sons, who are then affected (because they only have one X chromosome, whereas the women have two X chromosomes). Autosomal traits will affect females and males equally, (A) and (B). Sex-linked dominant diseases are rare and could not skip generations—rather, any woman who had an X chromosome with the trait for Disease A would be afflicted with Disease A (C).

11. **C**   Transduction is the process by which DNA can be transferred laterally from one bacterial cell to another by a virus. Mitosis and meiosis do not involve exchange of genetic information; furthermore, meiosis does not occur in bacterial cells (A, B). Transfection refers to the process of eukaryotic cells taking up free DNA from the environment (D).

12. **C**   Mutations occur when genetic information is improperly copied and then not corrected; while viruses with DNA genomes use host cell machinery including replication enzymes that have error-checking mechanisms, there are not error-checking mechanisms for RNA synthesis, so viruses with RNA genomes have no way for their genome to be "proofread". The types of genome of a virus does not dictate the speed at which it goes through a cycle; rather, the actual genes themselves would influence this, (A) and (B). As mentioned, DNA viruses use host cell machinery to ensure they have a mechanism for checking errors in replication (D).

13. **A**   If acetylcholine esterase was not present in the synaptic cleft, there would be more acetylcholine around for longer because it would not be broken down, which would then continue to stimulate the post-synaptic neuron to fire more action potentials. Neurotransmitter in the synaptic cleft sends its signal to the post-synaptic (not the pre-synaptic) neuron so eliminate choice (C) and (D). If acetylcholine was inhibitory, then there would be fewer action potentials fired by the post-synaptic neuron (B).

14. **C**   DNA is always synthesized 5' → 3', and each strand of the parent DNA is used as a template for a complimentary strand of daughter DNA. In eukaryotes, DNA synthesis occurs in the nucleus, but prokaryotic cells lack a nucleus, so DNA synthesis occurs in the cytosol.

15. **C** The ratio of five white-haired to five black-haired baby mice is expected by Mendelian genetics if either black or white hair is the dominant trait, and the parent with the dominant trait is heterozygous; for example, if black hair is dominant over white hair, with the black-haired parent mouse being Bb and the white-haired parent mouse being bb, you would expect $\frac{1}{2}$ of their offspring to be Bb (black) and $\frac{1}{2}$ of their offspring to be bb (white). If each baby mouse had a different color of hair, this trait likely would not be under complete genetic control (A). A ratio of 9:1 is unexpected and not predicted by typical Mendelian genetics and thus does not provide the best support for the trait being controlled by genes alone, (B) and (D).

16. **D** Gentamicin-resistant *Acinetobacter* are growing on both areas III and IV; in area IV, only gentamicin-resistant bacteria is growing because it is grown in the presence of gentamicin, but because the bacteria in area III are have also been transformed, some of those also are gentamicin resistant. Wild type *Acinetobacter* would not be resistant to gentamicin.

17. **C** If ADH secretion was blocked, there would be less aquaporins in the collecting duct, and so less water would be able to be reabsorbed. Instead, more water would be left to go into the urine, making the urine more dilute (B). More water might be consumed in response to blockage of ADH secretion because less water could be reabsorbed and the person may feel dehydrated (A). There would not be any negative feedback from ADH if it could not be secreted (D).

18. **D** Growing follicles within the ovary produce most of the estrogen in human females, in response to FSH. LH and FSH come from the pituitary gland (A). The corpus luteum produces progesterone after the egg is ovulated (B). The endometrium does not produce any hormones but grows and sheds in response to hormone levels (C).

19. **B** mRNA processing occurs before the mRNA molecule leaves the nucleus. The other statements are all correct.

20. **D** Both the nervous system and the endocrine system control body activities to send messages to other parts of the body—the endocrine system relies on increasing and decreasing hormone levels and thus works more slowly than the nervous system, which relies on nerve impulses.

21. **D** Herd and schooling behaviors, as well as predator warnings, are all examples of cooperative behavior in animals which increase the fitness of the individual as well as survival of the population. A rabbit running from a predator will help that rabbit to survive, but it is not an example of a cooperative behavior.

Chapter 17
Big Idea 4 Drill 2

BIG IDEA 4 DRILL 2

Multiple-Choice Questions

1. Prokaryotes use methylation as a major mechanism of DNA protection because

    (A) methyl groups include bonds that are impossible to break

    (B) methyl groups physically block portions of DNA that enzymes would otherwise access to breakdown DNA

    (C) methyl groups are added to DNA by restriction enzymes in the cytoplasm

    (D) methyl groups prevent spontaneous breakdown of DNA that otherwise would occur in the cytoplasm

2. Which of the following structural properties is not shared by most lipids but is a property of phospholipids?

    (A) A chain of carbon atoms

    (B) Hydrogen bonding to some carbon atoms

    (C) Nonpolarity

    (D) Polar regions

3. Which of the following directional components of polymers are most similar with respect to polymerization?

    (A) The amino end of a protein and the 3' end of a DNA molecule

    (B) The amino end of a protein and the phosphorous end of a phospholipid

    (C) The carboxyl end of a protein and the 3' end of a DNA molecule

    (D) The carboxyl end of a protein and the 5' end of a DNA molecule

4. Which of the following is NOT a function of the endoplasmic reticulum?

    (A) Packaging of small molecules for transport in vesicles

    (B) Mechanical support for the cell

    (C) Protein synthesis via membrane-bound ribosomes

    (D) Compartmentalization of the cell

5. Which of the following cellular components increase surface area available for ATP production?

(A) The outer membrane of mitochondria
(B) The inner membrane of mitochondria
(C) The membrane of the endoplasmic reticulum
(D) The cisternae of the Golgi complex

6. Which of the following accurately describes the ways in which energy and matter move within an ecosystem?

(A) Energy flows, while matter is recycled
(B) Energy is recycled, while matter flows
(C) Energy and matter both flow
(D) Energy and matter are both recycled

7. How do humans most commonly impact ecosystems and life on Earth?

(A) Accelerate change
(B) Decelerate change
(C) Help to avoid any change
(D) Have no impact on change

8. A food web for a forest habitat that spans 30 km² is shown below. The biomass of the primary producers is consistently distributed throughout the forest and totals 1,250 kg/km². A new housing development is being built which will permanently reduce the biomass of the primary producers by 75%, and will remove all hedgehogs and squirrels. Which of the following will most likely result from completion of the housing development?

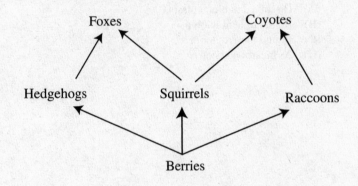

(A) The biomass of foxes will be significantly reduced.
(B) The biomass of raccoons will be 12 kg, and the biomass of coyotes will be 24 kg.
(C) The biomass of coyotes will be significantly increased.
(D) The number of raccoons will decrease by 75%, and the number of coyotes will decrease by 85%.

9. A point mutation that results in a codon for an amino acid with a hydrophobic R-group replacing a codon for an amino acid with a hydrophilic R-group on the exterior of the translated protein will most likely cause a change in

(A) DNA structure due to abnormal base-pairing
(B) mRNA structure due to abnormal bonds between nucleotides
(C) protein secondary structure due to abnormal hydrophobic interactions between R-groups in the backbone of the molecule
(D) molecular properties of the protein due to abnormal interactions between the protein and other molecules

10. How is oxygen primary transported in the body?

(A) Dissolved in blood plasma
(B) Bound to hemoglobin
(C) Bound to myoglobin
(D) As bicarbonate ions

11. Which of the following processes results in the greatest number of functional daughter cells?

(A) Mitosis
(B) Oogenesis
(C) Spermatogenesis
(D) All of the above processes result in an equal number of daughter cells.

12. When white mice are mated with black mice, all of their offspring are grey mice. This is most likely an example of

(A) classical dominance
(B) codominance
(C) incomplete dominance
(D) polygenetic traits

13. The figure below depicts the hemoglobin saturation curve, which indicates how much hemoglobin is bound by oxygen in different conditions. The solid black line shows normal body conditions, while the dotted line and the conditions listed on the right represent hemoglobin saturation during exercise.

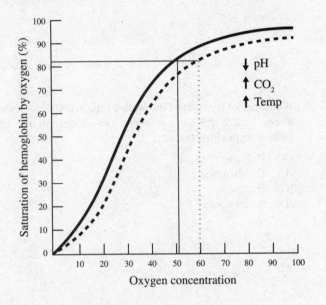

According to the graph, in which situation would hemoglobin be least saturated by oxygen molecules?

(A) Non-exercise and an oxygen concentration of 80%
(B) Non-exercise and an oxygen concentration of 70%
(C) Exercise and an oxygen concentration of 80%
(D) Exercise and an oxygen concentration of 70%

14. A urinary tract infection is an ascending infection of bacteria in the urinary system, from the external to the internal environment. Sometimes, the infection can then ascend even further up into the kidney, causing a kidney infection. Which best describes the route that bacteria must traverse from the external environment to the kidney to cause such an infection?

(A) Ureter → Bladder → Urethra → Kidney
(B) Urethra → Ureter → Bladder → Kidney
(C) Bladder → Ureter → Urethra → Kidney
(D) Urethra → Bladder → Ureter → Kidney

15. Which part of the nervous system is NOT involved in involuntary body activities?

    (A) Autonomic nervous system
    (B) Parasympathetic nervous system
    (C) Somatic nervous system
    (D) Sympathetic nervous system

16. Which of the following is the best description of the relative refractory period?

    (A) The time in which a second action potential cannot be fired due to inactivation of sodium channels
    (B) The time in which a second action potential cannot be fired due to inactivation of potassium channels
    (C) The time in which a second action potential can be fired but is more difficult than normal to fire, due to hyperpolarization of the neuron
    (D) The time in which a second action potential can be fired but is more difficult than normal to fire, due to depolarization of the neuron

17. Which of the following is NOT an accessory organ of the human digestive system?

    (A) Appendix
    (B) Gall bladder
    (C) Liver
    (D) Salivary glands

18. What special function of the walls of the digestive system allows food to move in a wavelike motion through the gastrointestinal tract?

    (A) Bolusing
    (B) Emulsification
    (C) Peristalsis
    (D) Reabsorption

19. By what process are oxygen and carbon dioxide exchanged across the alveolus and capillary in the lungs?

    (A) Facilitated diffusion
    (B) Passive diffusion
    (C) Primary active transport
    (D) Secondary active transport

20. Which of the following is most likely to occur during exercise?

    (A) Blood pH decreases, causing chemoreceptors to send signals to increase respiratory rate
    (B) Blood pH decreases, causing chemoreceptors to send signals to decrease respiratory rate
    (C) Blood pH increases, causing chemoreceptors to send signals to increase respiratory rate
    (D) Blood pH increases, causing chemoreceptors to send signals to decrease respiratory rate

21. Which of the following accurately describes the path of blood through the systemic circulation?

    (A) Blood leaves the heart from the right ventricle through the aorta, travels through arteries to arterioles, then to capillaries, then through venules to veins, and then enters the heart from the superior and inferior vena cava into the right ventricle.
    (B) Blood leaves the heart from the left atrium through the aorta, travels through arteries to arterioles, then to capillaries, then through venules to veins, and then enters the heart from the superior and inferior vena cava into the right ventricle.
    (C) Blood leaves the heart from the left ventricle through the aorta, travels through arteries to arterioles, then to capillaries, then through venules to veins, and then enters the heart from the superior and inferior vena cava into the right atrium.
    (D) Blood leaves the heart from the right atrium through the aorta, travels through arteries to arterioles, then to capillaries, then through venules to veins, and then enters the heart from the superior and inferior vena cava into the right atrium.

22. Gastric juices have a pH of approximately 1.5. Compared with blood which has a pH of 7.5, gastric juices are

    (A) more acidic because they have more H^+ ions
    (B) more acidic because they have more OH^- ions
    (C) more basic because they have more H^+ ions
    (D) more basic because they have more OH^- ions

23. Digestion of proteins doesn't mostly occur until it reaches which of the following structures of the gastrointestinal (GI) tract?

   (A) Mouth
   (B) Esophagus
   (C) Stomach
   (D) Small Intestine

24. Bile salts are produced by the liver and excreted via the gall bladder into the duodenum. One of the primary roles of bile salts is to neutralize the acidic chyme released by the stomach. Which of the following is the other primary role of bile?

   (A) To regulate the release of cholecystokinin (CCK), a digestive hormone
   (B) To promote beneficial bacterial growth in the large intestine
   (C) To cleave proteins for absorption
   (D) To emulsify lipids

25. Erythrocytes (red blood cells) are unique cells which makeup the majority of the solid elements of blood. Which of the following is NOT a characteristic of erythrocytes?

   (A) They have a nucleus.
   (B) They carry oxygen bound to hemoglobin.
   (C) They are derived from bone marrow.
   (D) They have a biconcave disk shape.

Chapter 18
Big Idea 4
Drill 2 Answers and
Explanations

ANSWER KEY

1. B
2. D
3. C
4. A
5. B
6. A
7. A
8. A
9. D
10. B
11. C
12. C
13. D
14. D
15. C
16. C
17. A
18. C
19. B
20. A
21. C
22. A
23. C
24. D
25. A

ANSWERS AND EXPLANATIONS

Multiple-Choice Questions

1. **B** Because the DNA of prokaryotes is located in the cytoplasm, it is important that this DNA is methylated to prevent enzymes recognizing and binding to restriction sites in the DNA—these enzymes are designed to breakdown DNA, and are in the cytoplasm of prokaryotes as a protective mechanism to breakdown viral DNA that may infect the cell. Physical blockage of the restriction sites protects DNA that is methylated, not the strength of the bonds within the methyl groups (A). Methyl groups protect against restriction enzymes; they are not added by these enzymes but rather specific methylation enzymes (C). DNA would not spontaneously breakdown in the cytoplasm but is broken down by restriction enzymes in the cytoplasm (D).

2. **D** Phospholipids, unlike most lipids, have polar head groups that interact with other polar molecules like water. Chains of carbon atoms with varying levels of saturation, or hydrogen bonds, and non-polarity are all common properties to most lipids.

3. **C** Amino acids are added to the carboxyl end of the growing peptide chain in protein synthesis, and nucleotides are added to the 3' end of the growing strand in DNA synthesis. The 5' end of the DNA strand and the amino end of the peptide chain are opposite of the end where new molecules are added (A, D). Phospholipids do not have directionality in the same way that DNA molecules and proteins do during synthesis (B).

4. **A** The Golgi complex is the site of packaging of materials into vesicles for transport, not the endoplasmic reticulum. The endoplasmic reticulum does provide mechanical support and compartmentalization and is a site of protein synthesis via the ribosomes found on its membrane.

5. **B** The inner membrane of the mitochondria is folded into cristae, which contain enzymes for ATP production and increase the surface area available for ATP production. The other membranes mentioned are not utilized for ATP production. The cisternae of the Golgi complex are flattened membrane sacs, and are involved of synthesis and packing of materials in vesicles as well as lysosome production.

6. **A** Energy flows up through levels of an ecosystem, with some lost in the transition from one level to the next, while matter is recycled.

7. **A** Human impact on ecosystems and life on Earth tends to accelerate change at both local and global levels. It is unlikely that humans would avoid change or have no impact on change, and it is nearly impossible that humans would decelerate change in an ecosystem.

8. **A** If the biomass of the primary producers in this food web, the plants making the berries, is dramatically reduced, as well as the hedgehogs and squirrels, which the foxes eat, the biomass of foxes will also be significantly reduced. It is not possible to calculate specific biomasses or percentage reductions of different members of the food web with the information given. There is no evidence that the biomass of coyotes would be increased; in fact, this would likely decrease because squirrels are being removed and the berries which the raccoons eat are being reduced (B).

9. **D** Because there is a change in the amino acid (from hydrophobic to hydrophilic) on the exterior of the protein, this is most likely to change how it interacts with other molecules due to its position on the external portion of the protein. Base-pairing in the DNA molecule and nucleotide bonds in the mRNA should not be altered by a point mutation. Protein secondary structure is unlikely to be affected because the altered amino acid is on the exterior of the protein and thus is not interacting with other amino acids within the protein.

10. **B** Oxygen is primarily transported through the body bound to hemoglobin. A small amount is dissolved in the plasma, and some is bound to myoglobin, but these are not the major mechanisms for hemoglobin transport. Carbon dioxide is primarily transported through the body as bicarbonate ions.

11. **C** Spermatogenesis, male gamete production via meiosis, results in four sperms from one original spermatocyte. Mitosis results in two identical daughter cells (A). Oogenesis, female gamete production, results in one functional ovum (egg) as well as three polar bodies which disintegrate (B).

12. **C** An intermediate phenotype (grey) resulting from two different phenotypes (white and black) is indicative of incomplete dominance. Classical dominance refers to one phenotype (and gene) being dominant over another (A). Codominance occurs when two alleles are equally and independently expressed (B). A polygenetic traits is one that is determined by more than one gene (D).

13. **D** Looking at the graph, the saturation of hemoglobin by oxygen is the lowest under the conditions of exercise and an oxygen concentration of 70%. At all of the other conditions, the graph shows a higher saturation of hemoglobin by oxygen.

14. **D** A urinary tract infection occurs when bacteria enters through the ureter and travels to the bladder; from there, bacteria can travel up through the ureter to the kidney.

15. **C** The somatic nervous system controls skeletal muscle movement, which is under voluntary control. The autonomic nervous system controls involuntary activities and is further subdivided into the parasympathetic and sympathetic nervous systems.

16. **C** The relative refractory period occurs after sodium channels are reset and thus can open, but the neuron is hyperpolarized (more negative than resting potential) and thus it is more difficult than normal to fire an action potential. Depolarization occurs during an action potential (D). The time in which a second action potential cannot be fired due to inactivation of sodium channels is referred to the absolute refractory period (A). Potassium channels close but are never inactivated (B).

17. **A** Although the appendix is connected to the intestine, it is not an accessory organ of the digestive system, as it does not have any digestive functions. The gall bladder, liver, and salivary glands all aid in digestion and thus are accessory organs of the digestive system.

18. **C** Peristalsis results from specialized contraction of the muscles in the walls of the digestive tract, allowing food to move down the gastrointestinal tract in a wavelike motion. A bolus is formed in the mouth (A). Bile is made in the liver and stored in the gall bladder and emulsifies fats (B). Reabsorption of water and salts occurs in the large intestine (D).

19. **B** Oxygen and carbon dioxide are able to move between the capillary and the alveolus of the lungs via passive diffusion. Because they are moving along their concentration gradients and do not need carrier proteins, this process is not active transport or facilitated diffusion.

20. **A** During exercise, lactic acid is produced, causing a decrease in blood pH, which then triggers chemoreceptors to send signals to the diaphragm and intercostal muscles increase the respiratory rate.

21. **C** Blood leaves the heart from the left ventricle, travels through the systemic circulation, and then re-enters the heart into the right atrium.

22. **A** Gastric juices are much more acidic than blood because of the release of hydrochloric acid which produces H^+ ions in solution.

23. **C** Proteins are not largely digested until they are immersed in the gastric juices, which contain hydrochloric acid (low pH) and are enzymatically digested by pepsin. Although a small amount of the proteins may be disrupted by the chemical and mechanical activity of the mouth, these actions are negligible compared with the activity of the stomach.

24. **D** The primary roles of bile are to emulsify lipids and to neutralize chyme as it enters the duodenum of the small intestine.

25. **A** Erythrocytes (red blood cells) are unique because they have no nucleus. They are designed to carry as much hemoglobin, for the purpose of carrying oxygen, as possible. They are derived from bone marrow and have a biconcave disk shape.

Chapter 19
Big Idea 1 Drill 3

BIG IDEA 1 DRILL 3

Multiple-Choice Questions

1. Which of the following statements correctly articulates the Hardy-Weinberg principle?

    (A) Inbreeding within a population causes an increase in homozygosity for some genes.

    (B) Allele and genotype frequencies in a population will remain constant from generation to generation in the absence of evolutionary pressures.

    (C) The natural environment of an organism selects for traits that confer a reproductive advantage, causing gradual changes from generation to generation.

    (D) New alleles are created through random changes in the DNA sequence.

2. Which of the following is an example of disruptive selection?

    (A) Fossil records show that the size of black bears in Europe decreased during interglacial periods of the ice ages.

    (B) Babies with medium birth weights survive more often than do babies with either low or high birth weights.

    (C) The beak size of a population of finches increases in response to the abundance of a particular type of seed for food.

    (D) The gene coding for the color of rabbit fur exhibits incomplete dominance, and rabbits with white fur or black fur have a survival advantage over rabbits with grey fur.

3. Which of the following is an example of convergent evolution?

    (A) The capacity for flight has evolved independently in flying insects, birds, and bats.

    (B) Over 80 varieties of finches evolved from one species.

    (C) The American red wolf is a hybrid species between grey wolf and coyote.

    (D) A new species of fruit flies emerged in North America after the introduction of a non-native fruit.

4. Each of the following is a type of symbiotic relationship EXCEPT

    (A) mutualism

    (B) polymorphism

    (C) commensalism

    (D) parasitism

5. In pea plants, the tall allele (T) for the height gene is dominant, and the short allele (t) is recessive. The green allele (G) for the pea pod color gene is dominant, and the yellow allele (g) is recessive. The height gene and the pea pod color gene assort independently. Assume that a cross is performed between two pea plants. The first parental plant is homozygous dominant for the height gene (TT) and heterozygous for the pea pod color gene (Gg). The second parental plant is heterozygous for the height gene (Tt) and homozygous recessive for pea pod color gene (gg). What is the likelihood that an offspring of the cross would be tall with yellow pea pods?

(A) 0%
(B) 25%
(C) 50%
(D) 75%

6. Some bacterial cells are surrounded by two separate lipid membranes with a thin layer of peptidoglycan between the membranes. Which of the following is true regarding this type of bacteria?

(A) This type of bacteria is gram negative because the gram stain cannot bind to the peptidoglycan.
(B) This type of bacteria is gram positive because the gram stain binds to the peptidoglycan.
(C) This type of bacteria is gram negative because the stain binds to the outermost membrane.
(D) This type of bacteria is gram positive because the stain cannot bind to the outermost membrane.

7. Suppose a population is in genetic equilibrium and the dominant allele of a gene represents 20% of the alleles in the population. What percentage of the population is homozygous dominant?

(A) 4%
(B) 20%
(C) 64%
(D) 80%

8. Which of the following describes the process of double fertilization in flowering plants?

(A) A pollen grain divides into two sperm nuclei, and each sperm nucleus fuses with an egg nucleus. The result is the production of two zygotes.

(B) A pollen grain divides into two sperm nuclei, and one sperm nucleus fuses with an egg nucleus while the other sperm nucleus fuses with two polar nuclei. The result is the production of one zygote and one endosperm.

(C) A pollen grain divides into two sperm nuclei, and both sperm nuclei fuse with the same egg nucleus. The result is the production of one endosperm.

(D) A pollen grain fuses with an egg, and the fertilized cell divides into two zygotes.

9. Which of the following terms refers to a community of living organisms and the nonliving components of their environment?

(A) Population
(B) Species
(C) Habitat
(D) Ecosystem

10. Which of the following distinguishes an animal cell from a plant cell?

(A) An animal cell contains a nucleus, whereas a plant cell does not.

(B) An animal cells is surrounded by a cell wall, whereas a plant cell is not.

(C) A plant cell contains a nucleus, whereas an animal cell does not.

(D) A plant cell is surrounded by a cell wall, whereas an animal cell is not.

11. Which of the following is an example of exponential growth?

(A) The number of deer in a colony increases by six each year.

(B) The number of microorganisms in a culture doubles every 20 minutes.

(C) The number of humans on the Earth increases by one billion every 13 years.

(D) The number of birds in a colony is reduced to zero due to disease.

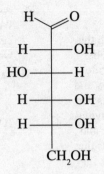

12. Which of the following molecules contains a carboxyl group?

(A)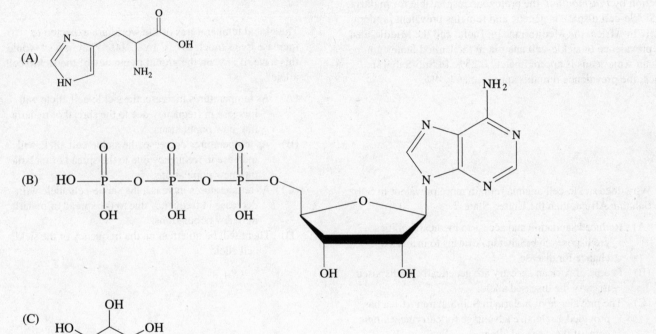

(B)

(C)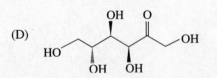

(D)

Questions 13-15

Sickle-cell anemia is a blood disorder associated with the formation of abnormal, rigid (sickle) shaped red blood cells. Sickled red blood cells lack flexibility and can lead to acute or chronic diseases including anemia, reduced blood flow and tissue damage, and cardiovascular disease. The disease is due to a single nucleotide mutation in the β-globin gene. Individuals homozygous for the mutant gene show severe disease and experience reduced life expectancy. However, individuals that are heterozygous show much less disease and experience very little impairment to life. Due to the drastic change in red blood cell architecture, sickle-cell anemia confers higher resistance to infection by *Plasmodium,* the protozoan responsible for malaria.

Sickle-cell disease is genetic and remains prevalent predominately in Africa, the Mediterranean, India, and the Middle East. The prevalence of sickle-cell anemia in the United States among African Americans is approximately 0.25%. In Sub-Saharan Africa, the prevalence remains approximately 4%.

13. Why does sickle-cell anemia remain more prevalent in Sub-Saharan Africa, than the United States?

    (A) Reduced sanitation and access to medical facilities predisposes Sub-Saharan Africans to increased chance for disease.
    (B) People of African ancestry are genetically predisposed to carry the diseased allele.
    (C) The prevalence of malaria in Sub-Saharan Africa has provided a selective advantage for carrying a single copy of the diseased allele.
    (D) Infection with malaria provides a selective disadvantage to individuals carrying the diseased allele.

14. Assume that in a sufficiently large population in Hardy-Weinberg equilibrium, the allelic frequency of the sickle-cell anemia allele is 0.1. What percentage of the population would be heterozygous for the disease?

    (A) 1%
    (B) 9%
    (C) 18%
    (D) 19%

15. The global temperatures of the world are expected to increase by as much as 6°C by 2,100. What impact would this have, if any, on the global frequency of the sickle-cell allele?

    (A) As temperatures increase, the sickle-cell allele will increase in frequency due to the spread of malaria into new populations.
    (B) As temperatures decrease, the sickle-cell allele will increase in frequency due to the spread of malaria into new populations.
    (C) As temperatures increase, the sickle-cell allele will decrease in frequency due to the spread of malaria into new populations.
    (D) There will be no effect on the frequency of the sickle-cell allele.

Questions 16-20 refer to the following information.

   Streptococcus pneumoniae is part of the normal flora of the upper respiratory tract. However, in susceptible individuals (particularly the elderly, immunocompromised, and the young) *S. pneumoniae* may cause pneumococcal diseases including pneumonia, bronchitis, and otitis media (ear infections). Traditionally, penicillin antibiotics were prescribed as a treatment for a *S. pneumoniae* infection. Penicillin was first discovered in 1928 and has been a critical antibiotic for the treatment of bacterial infections. However, recently, many species of bacteria have acquired resistance. The graph below displays the number of susceptible and resistant clinical isolates of *S. pneumoniae* between 1980 and 2000.

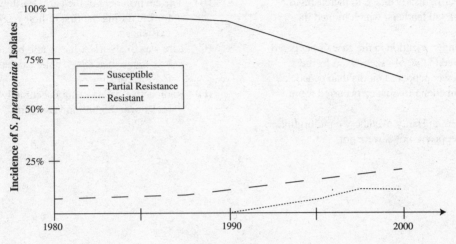

16. Approximately what percentage of clinical isolates surveyed in 1995 was identified as partially resistant to penicillin antibiotics?

   (A) 75%
   (B) 30%
   (C) 20%
   (D) 5%

17. Which of the following, if true, would NOT support the hypothesis that overuse of penicillin has selected for resistance in *S. pneumoniae* species?

   (A) The prescription of penicillin antibiotics has increased 30% between 1980 and 2000.
   (B) The average patient on penicillin antibiotics misses or fails to take their final two doses.
   (C) 23% of *S. pneumoniae* isolates in 2000 encode a penicillinase gene, which degrades penicillin, compared to only 2% in 1980.
   (D) A clinical survey published in 2005 showing that the incidence of susceptible *S. pneumoniae* had increased to 90% despite no change in prescription of penicillin antibiotics.

18. Suppose a new antibiotic was discovered which was functionally distinct but distantly related to the penicillin class of antibiotics. If physicians began to prescribe this antibiotic in lieu of penicillin antibiotics for *S. pneumoniae* infections, which of the following would you expect to occur?

   (A) The antibiotic would have limited efficacy due to pre-existent resistance to penicillin antibiotics.
   (B) The antibiotic would have high efficacy initially, however, *S. pneumoniae* would acquire increasingly higher resistance with time.
   (C) The antibiotic would have limited efficacy initially, however, *S. pneumoniae* would acquire increasingly higher susceptibility with time.
   (D) The antibiotic would have no effect, as *S. pneumoniae* is only susceptible to penicillin antibiotics.

19. The rates of evolution vary greatly depending upon the trait and species involved. The coloration of the peppered moth changed rapidly over the 200 years beginning with the Industrial Revolution, whereas the increase in size and color of the peacock tail took far longer over thousands of years. Why was the evolution of peppered moth coloration more rapid than the evolution of size and color of peacock tail feathers?

    (A) There was a selective disadvantage to increasing the size of peacock tail feathers, which limited its evolution.

    (B) There was a stronger selection in the case of peppered moths compared to that of peacock tail feathers.

    (C) There were far fewer peppered moths than peacocks, thus changes in allelic frequency occurred more quickly.

    (D) The peacocks were in Hardy-Weinberg equilibrium, whereas the peppered moths were not.

20. Based on multiple forms of scientific evidence, the Earth was formed approximately 4.6 billion years ago. However, evidence of life is not present in the fossil record until 3.5 billion years ago. Which of the following best explains why life cannot be detected during the first billion years of its existence?

    (A) Rocks dating before 3.5 billion years ago have yet to be found.

    (B) The environment of the early Earth was too hostile for life during the first billion years of Earth's existence.

    (C) Life was likely RNA-based and because RNA is far less stable than DNA, none can be detected in the fossil record.

    (D) Radioactive dating cannot accurately predict the age of rocks over 3.5 billion years old.

Questions 21-23 refer to the following diagram and paragraph

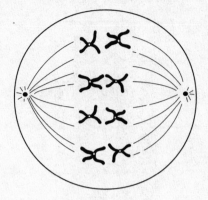

A geneticist is studying the impact of temperature on a lineage of cells acquired from an African green monkey. Upon dropping the temperature of the cells, she notices that all of the cells share a similar appearance (as shown above) and appear arrested in their cell cycle.

21. Which of the following may be the type of cell pictured above?

(A) An erythryocyte
(B) A hepatocyte
(C) A spermatocyte
(D) A muscle cell

22. During which phase of cell division is the cell above arrested?

(A) Prophase
(B) Metaphase I
(C) Anaphase
(D) Metaphase II

23. When the scientist rewarms the cells, they continue through their cell cycle and divide. By the time the cell above re-enters interphase, how many daughter cells will the cell above have produced?

(A) 1
(B) 2
(C) 4
(D) 8

24. Lamarckian evolution was one of the first widely accepted notion of evolution. However, Darwin's theory proved that Lamarck's view of evolution was wrong. Which of the following best describes the Lamarckian view of evolution?

(A) Organisms with traits that provide a survival or reproductive advantage are more likely to pass on their traits.

(B) Organisms that use helpful acquired traits will be able to pass them on to the next generation.

(C) Organisms that produce more offspring are more likely to pass on their traits than those that don't.

(D) Organisms that sustain more radiation will evolve more rapidly and adjust to new environments better than those that don't.

25. The molecular formula for glucose is $C_6H_{12}O_6$. What is the molecular weight of glucose in g/mol?

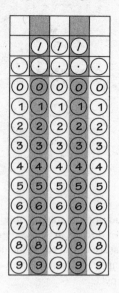

Chapter 20
Big Idea 1
Drill 3 Answers and Explanations

ANSWER KEY

1. B
2. D
3. A
4. B
5. C
6. A
7. A
8. B
9. D
10. D
11. B
12. A
13. C
14. C
15. A
16. C
17. D
18. B
19. B
20. B
21. C
22. B
23. C
24. B
25. 180

ANSWERS AND EXPLANATIONS

Multiple-Choice Questions

1. **B** The Hardy-Weinberg principle deals with population genetics, particularly under conditions that lack evolutionary pressures. Answer (B) correctly summarizes the Hardy-Weinberg principle. Answer (A) is incorrect because it is an example of non-random mating. The Hardy-Weinberg principle only applies when the population undergoes random mating. Answer (C) describes natural selection, so that answer can be eliminated. Answer (D) describes the phenomenon of mutation, so (D) is wrong. The Hardy-Weinberg principle only applies to high-specific conditions, which include no mutations.

2. **D** Disruptive selection favors organisms in a population with extreme traits. Answer (D) is an example of this because the extreme traits of black or white are favored over the blended trait of gray. Answer (D) is correct answer. Answers (A) and (C) are examples of directional selection. In directional selection, one extreme is favored over the other. For instance, small bears are favored over large bears, or large beaks are favored over small beaks. Thus, answers (A) and (C) are incorrect. Answer (B) is wrong because it is an example of stabilizing selection. In stabilizing selection, both extremes are disfavored. In this answer, medium birth weight is favored over the extremes of high and low birth weight.

3. **A** Convergent evolution refers to the evolution of similar traits in unrelated species. Answer (A) offers an example of this because three unrelated types of organisms acquired the common trait of flying. Answers (B) and (D) are incorrect because they are examples of divergent evolution, which is the accumulation of differences in related species. Answer (C) can be eliminated because it is an example of hybrid speciation, not convergent evolution.

4. **B** Answers (A), (C), and (D) are examples of symbiotic relationships, so those answers should be eliminated. Mutualism is a symbiotic relationship in which both organisms benefit from the relationship. Commensalism is a symbiotic relationship in which one organism lives off another with no harm to the host. Parasitism is a symbiotic relationship in which an organism harms its host. Polymorphism is not related to symbiosis, so answer (B) is correct. A polymorphism exists when there are two or more phenotypes for a gene in a population of animals.

5. **C** Because the two genes assort independently, each probability can be calculated separated. The parental genotypes for the height gene are TT and Tt. The offspring will be either homozygous dominant (TT) or heterozygous (Tt). Both of these genotypes correspond to a tall phenotype, so there is 100% likelihood that an offspring plant would have the tall phenotype. The parental genotypes for the pea pod color gene are Gg and gg. Half the offspring will be heterozygous (Gg),

and half the offspring will be homozygous recessive (gg). The Gg genotype corresponds to green pea pods, and the gg genotype corresponds to yellow pea pods. So 50% of the offspring will have the yellow pea pod allele.

The probability of two independent events equals the product of the probabilities of each event. The two probabilities in this question are 100% (1) and 50% (0.5).

$$1 \times 0.5 = 0.5$$

Because 100% times 50% equals 50%, the answer is (C).

6. **A** Gram staining is a technique used to categorize bacteria into gram positive and gram negative. Bacterial cells with an exposed layer of peptidoglycan are gram positive because the Gram stain binds to the peptidoglycan. As the questions states, however, there are some bacterial cells with an extra lipid membrane that surrounds the peptidoglycan. The Gram stain cannot penetrate the outermost membrane, so the stain does not bind to the peptidoglycan, and these bacteria are gram negative. Thus, answer (A) is correct. Answer (B) is incorrect because the stain is unable to penetrate the outermost membrane, so the stain cannot bind to the peptidoglycan. Although the bacteria in question are gram negative, answer (C) is wrong because the Gram stain does not bind to the outermost membrane. Answer (D) is wrong because the bacteria in question are gram negative, not gram positive; the relevant consideration is whether the stain binds to the peptidoglycan.

7. **A** The question states that the population is in genetic equilibrium, which indicates that the Hardy-Weinberg principles applies. Thus, in this population, the gene pool is stable. In the Hardy-Weinberg equations, p represents the frequency of the dominant allele, in this case 20% or 0.2. The homozygous dominants are represented by p^2, which in this case is 0.2^2. That equals 0.04 or 4%, so answer (A) is correct. Answer (B) simply mimics the figure given in the problem, so answer (B) is incorrect. Answer (D) corresponds to q in the Hardy-Weinberg equations. Because the question asks for p^2, not q, answer (D) is wrong. Answer (C) provides the frequency of the homozygous *recessive*, q^2, not the homozygous dominant. Therefore, answer (C) is wrong.

8. **B** Flowering plants carry out a process called double fertilization. In this process, the pollen grain divides into two sperm nuclei before fusing with the egg nucleus, so answer (D) is incorrect. Only one sperm nucleus will fuse with an egg nucleus to form a plant, so answers (A) and (C) are wrong. The other sperm nucleus fuses with two polar nuclei to form an endosperm. The endosperm provides food for the plant embryo. Thus, answer (B) is correct.

9. **D** A population is a group of interbreeding organisms, but it does not include the nonliving components of the environment, so answer (A) is wrong. A species is often defined as a group of organisms capable of interbreeding and producing fertile offspring. The definition of species does not include the nonliving components of the environment, so answer (B) is incorrect. A habitat is an environmental area that is inhabited by a particular species, but the definition of habitat does not include the organisms, so answer (C) is wrong. The definition of ecosystem corresponds to the question, so answer (D) is correct.

10. **D** One distinction between an animal cell and a plant cell is that a plant cell is surrounded by a cell wall, whereas an animal cell is not. That corresponds to answer (D). Answers (A) and (C) are incorrect because plant cells and animals cell contain nuclei. Answer (B) is incorrect because it reverses the correct relationship: Plant cells are surrounded by cell walls (not animal cells, as stated in the answer choice).

11. **B** Exponential growth refers to a growth rate that *increases* in proportion to its current value. In answer (B), the number of microorganisms will grow at an increasing rate. For example, if the culture originally contained one microorganism, the number would increase to two after 20 minutes, and then to four after another 20 minutes. In the first 20 minutes, the number of microorganisms increased by 1, and in the second 20 minutes, the number increased by 2. This is an example of an exponentially increasing growth rate, so (B) is the correct answer. In answer (A), the number of deer increases by six each year regardless of how many deer are in the community and this growth is linear. Linear growth is also the case in choice (C). Answer (D) is an example of decay, not growth, so answer (D) can be eliminated.

12. **A** A carboxyl group (–COOH) consists of a carbon bonded to two oxygen atoms. One of the oxygen atoms is double bonded the carbon, and the oxygen is also bonded to hydrogen. The molecule in answer (A) contains a carbon atom in the upper right portion of the diagram that is doubled bonded to oxygen and then single bonded to another oxygen which is in turn bonded to hydrogen. That is the carboxyl group, so answer (A) is correct. Although there are many oxygen atoms in answers (B), (C), and (D), none of the carbon atoms in those answers choices is bonded to *two* separate oxygen atoms. Thus, answers (B), (C), and (D) are incorrect.

13. **C** The presence of sickle-cell anemia confers a selective advantage to surviving malaria infection due to resistance to the disease (eliminating D). Because sickle-cell disease is due to genetic factors rather than a microbial infection, sanitation and access to medical facilities (A) has very little impact on the prevalence of the disease. Sickle-cell disease is more prevalent in populations located in tropical and sub-tropical locations. The prevalence of the disease is not tied to specific ethnicity (B), note that African Americans (whom have African ancestry), have a significantly lower incidence of sickle-cell anemia than Sub-Saharan Africans.

14. **C** The allelic frequency of the diseased β-globin gene (q) is 0.1, which means that the allelic frequency of the normal β-globin gene (p) is 0.9 (1 − 0.1). Because there are two different combinations of alleles, which result in a heterozygous genotype (pq or qp), than the frequency of heterozygotes will be equal to $2pq$ ($2 \times 0.1 \times 0.9$) or 0.18.

15. **A** Increases in global temperatures will result in increased spread of the mosquito vectors, which carry the malaria gene. Because malaria selects for increased prevalence of the sickle-cell allele, the incidence of sickle-cell disease will most likely increase.

16. **C**   Using the figure provided, the isolates that were partially resistant (long dash line) represented approximately 20% of the surveyed isolates in 1995.

17. **D**   A survey showing that the incidence of susceptible *S. pneumoniae* had increased to 90% despite no change in prescription of penicillin antibiotics would suggest that exposure and use of antibiotics is not tied to overuse but some other factor of infection. Increased prescription of antibiotics (A) would support the hypothesis of overuse. Patients failing to finish antibiotic regimens would permit some bacteria to survive treatment and potentially develop resistance (B). The increased presence of penicillinase genes (C) suggests that the bacteria are acquiring resistance specifically to penicillin antibiotics.

18. **B**   The new antibiotic would likely follow a similar pattern to penicillin. Initially, the drug would maintain high efficacy, however over time and exposure, *S. pneumoniae* would be expected to acquire increasingly higher resistance.

19. **B**   The more rapid selection of color for peppered moths was due to higher selective pressure than in the case of peacock tail feathers. During the Industrial Revolution, ash and soot covered most surfaces. Darker moths were able to blend in far better than lighter moths and thus avoid predation. The strength of the selection rapidly shifted the prevalence of the color bias. In the case of peacock tail feathers, the sexual selection exhibited far less pressure, thus requiring longer time to change allelic frequencies. There were likely far more moths than peacocks (C) and neither population was truly in Hardy-Weinberg equilibrium since they were undergoing natural selection and non-random mating (D).

20. **B**   Based on scientific evidence, the environment of the early Earth was not conducive to life until approximately 3.9 billion years ago. Rocks far older than 3.5 billion years have been discovered (A), including a zircon crystal in Australia dating 4.4 billion years old. DNA and RNA degrade over time and are not used for age determination of fossil samples (C). Radioactive dating can and has been used for predicting the age of rocks older than 3.5 billion years old (D).

21. **C**   The cell shown is undergoing meiosis rather than mitosis as evident by the alignment of the chromosomes in the cell center during metaphase. Only gametic cells undergo meiosis. Erythrocytes (red blood cells; A), hepatocytes (liver cells; B), and muscle cells (D), are all somatic cells and undergo mitosis during cell division.

22. **B**   The cell is arrested in metaphase I because the homologous chromosomes are paired in the center of the cell. During metaphase II, the chromatids are separated.

23. **C**   During spermatogenesis, one spermatocyte gives rise to four spermatozoans by the completion of meiosis.

24. **B**   Lamarck believed that acquired traits could be passed on from generation to generation.

25. **180**   Each glucose molecule consists of six carbon atoms, six oxygen atoms, and 12 hydrogen atoms. The atomic weights for carbon, oxygen, and hydrogen are 12, 16, and 1 respectively. To calculate the molecular weight, the number of atoms of each type is multiplied by the atomic number of that type, and the products are then added together. The calculation for this problem is:

(6 carbon atoms $\times$ 12) + (6 oxygen atoms $\times$ 16) + (12 hydrogen atoms $\times$ 1) = 180 g/mol.

Chapter 21
Big Idea 2 Drill 3

BIG IDEA 2 DRILL 3

Multiple-Choice Questions

1. Phospholipids are amphipathic because

   (A) the fatty acid tails are attracted to positive charges while the phosphate head is attracted to negative charges

   (B) phospholipids are composed of carbon, hydrogen, oxygen, and phosphorous atoms

   (C) the phosphate head is hydrophilic while the fatty acid tails are hydrophobic

   (D) the fatty acid tails can be either saturated or unsaturated

2. Which of the following describes the bond between glycerol and a fatty acid in a triglyceride molecule?

   (A) Ester linkage
   (B) Peptide bond
   (C) Ionic bond
   (D) Hydrogen bond

3. If Solution X has a pH of 2 and Solution Y has a pH of 4, which of the following is true?

   (A) Solution X is more acidic than Solution Y, and Solution Y has twice as many hydrogen ions as Solution X.

   (B) Solution Y is more acidic than Solution X, and Solution Y has twice as many hydrogen ions as Solution Y.

   (C) Solution X is more acidic than Solution Y, and Solution X has 100 times as many hydrogen ions as Solution Y.

   (D) Solution X is more acidic than Solution Y, and Solution Y has 100 times as many hydrogen ions as Solution X.

4. Which of the following is a polysaccharide that serves as a form of energy storage in the liver and muscles of humans?

   (A) Glucose
   (B) Sucrose
   (C) Arginine
   (D) Glycogen

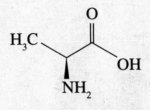

5. The molecule alanine is diagramed above. Each of the following is a functional group of alanine EXCEPT

(A) carboxyl group
(B) phenyl group
(C) methyl group
(D) amino group

6. Insulin is a peptide hormone secreted by cells in the pancreas. Which of the following organelles is the site of insulin synthesis within the pancreatic cells?

(A) Rough endoplasmic reticulum
(B) Smooth endoplasmic reticulum
(C) Lysosomes
(D) Mitochondria

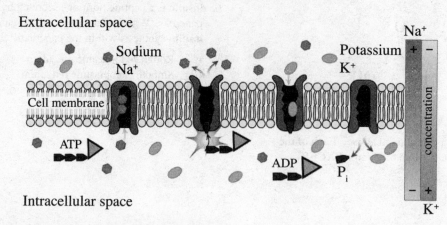

Extracellular space

Sodium
Na⁺

Potassium
K⁺

Na⁺

Cell membrane

concentration

ATP

ADP

P$_i$

K⁺

Intracellular space

7. The sodium-potassium pump is an enzyme located in
    the plasma membrane of all animal cells. The sodium
    potassium pump is illustrated above in four different stages
    of its mechanism. The pump uses energy derived from ATP
    hydrolysis to move three sodium ions out of the cell and two
    potassium ions into the cell. The movement of sodium and
    potassium ions across the plasma membrane in this fashion is
    an example of which of the following?

    (A) Passive transport
    (B) Active transport
    (C) Facilitated transport
    (D) Endocytosis

8. Which of the following molecules is synthesized in the mitochondria?

(A) Hydrogen peroxide
(B) Glucose
(C) Triglyceride
(D) Adenosine triphosphate

10. The DNA genome of *Escherichia coli* is located within which component of the cell?

(A) Mitochondrion
(B) Nucleus
(C) Cytosol
(D) Transmembrane protein

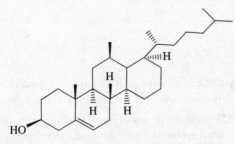

9. A cholesterol molecule is diagrammed above. Each of the following statements regarding cholesterol is true EXCEPT

(A) cholesterol is an alcohol
(B) cholesterol plays a role in stabilizing membrane fluidity
(C) cholesterol is a polypeptide found in the phospholipid bilayer
(D) there are 27 carbon atoms in a cholesterol molecule

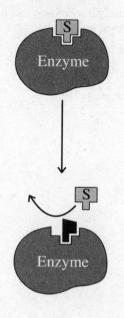

12. What is the first step of the Krebs cycle?

   (A) Glucose is phosphorylated.
   (B) Pyruvate is converted to acetyl CoA.
   (C) Ubiquinone accepts two electrons from $FADH_2$.
   (D) Acetyl CoA combines with oxaloacetate to form citrate.

13. ATP synthase is an enzyme found in mitochondria. Which of the following describes the function ATP synthase?

   (A) ATP synthase phosphorylates ADP using energy from protons moving down an electrochemical gradient.
   (B) ATP synthase hydrolyzes ATP which releases energy in the form of heat.
   (C) ATP synthase pumps protons from the matrix to the intermembrane space using energy from the hydrolysis of ATP.
   (D) ATP synthase accepts and donates electrons as part of the electron transport chain.

11. The diagram above depicts an enzyme, substrate S, and inhibitor I. How are these cellular molecules related to each other?

   (A) Inhibitor I decreases the rate of a reaction involving substrate S through allosteric inhibition.
   (B) Inhibitor I increases the rate of a reaction involving substrate S through allosteric inhibition.
   (C) Inhibitor I decreases the rate of a reaction involving substrate S through competitive inhibition.
   (D) Inhibitor I increases the rate of a reaction involving substrate S through competitive inhibition.

14. How many ATP molecules are consumed and produced in the glycolysis reaction of one glucose molecule?

(A) One ATP molecule is consumed and two ATP molecules are produced for a net increase of one ATP molecule.

(B) No ATP molecules are consumed and two ATP molecules are produced for a net increase of two ATP molecules.

(C) Two ATP molecules are consumed and four ATP molecules are produced for a net increase of two ATP molecules.

(D) No ATP molecules are consumed and four ATP molecules are produced for a net increase of four ATP molecules.

15. Each of the following occurs in muscle cells during periods of strenuous exercise EXCEPT

(A) an oxygen debt occurs.
(B) glucose is converted to glycogen.
(C) the concentration of NADH and $FADH_2$ increases.
(D) anaerobic respiration produces lactic acid.

16. The conversion of carbon dioxide to 3-phosphoglycerate occurs in which of the following chemical pathways?

(A) The Krebs cycle
(B) CAM pathway
(C) C_4 pathway
(D) C_3 pathway

17. Which of the following best describes photosystem II of plant cells?

(A) P680 chlorophyll absorbs light and passes excited electrons down an electron transport chain to produce ATP.

(B) P680 chlorophyll absorbs light and passes excited electrons down an electron transport chain to produce NADPH.

(C) P700 chlorophyll absorbs light and passes excited electrons down an electron transport chain to produce ATP.

(D) P700 chlorophyll absorbs light and passes excited electrons down an electron transport chain to produce NADPH.

18. PEP carboxylase is an enzyme used for carbon fixation in C_4 plants. Under certain conditions, the C_4 pathway is more efficient than the C_3 pathway. Each of the following statements explains the efficiency of the C_4 pathway EXCEPT

(A) as compared to rubisco, PEP carboxylase has a higher affinity for CO_2, which reduces photorespiration
(B) rubisco catalyzes the formation of an unstable six-carbon intermediate, which quickly decays into two three-carbon molecules
(C) in bundle-sheath cells of C_4 plants, rubisco is isolated from atmospheric oxygen
(D) in bundle-sheath cells of C_4 plants, rubisco is exposed to high concentrations of CO_2 released by decarboxylation of four-carbon molecules

19. The light-dependent reactions occur in which of the following organelles of a plant cell?

(A) Vacuole
(B) Cell wall
(C) Chloroplast
(D) Cytoplasm

20. Which of the following cellular processes occurs in the thylakoids?

(A) The Krebs cycle
(B) The light-dependent reactions
(C) The light-independent reactions
(D) The C_3 cycle

21. Bacterial cell walls are composed of which of the following polymers?

(A) Cellulose
(B) Chitin
(C) Silica
(D) Peptidoglycan

22. The bacterium *Staphylococcus* is a facultative anaerobe. What will happen if a laboratory biologist attempts to culture *Staphylococcus* cells on a nutrient-rich medium in the presence of atmospheric oxygen?

(A) The cells will die because oxygen is toxic to anaerobes.
(B) The cells will survive, but metabolism cannot occur in the presence of oxygen.
(C) The cells will survive and utilize fermentation for metabolism.
(D) The cells will survive and utilize aerobic respiration for metabolism.

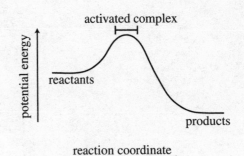

24. The reaction depicted above is

(A) exothermic because the reactants have less energy than the products.
(B) endothermic because the products have less energy than the reactants.
(C) exothermic because the products have less energy than the reactants.
(D) endothermic because the reactants have less energy than the products.

23. Which of the following types of transport involves the movement of a substance from an area of higher concentration to an area of lower concentration?

(A) Active transport
(B) Facilitated transport
(C) Sodium-potassium pump
(D) Antiport

25. What is the net ATP production from the glycolysis of one glucose molecule?

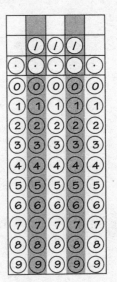

Chapter 22
Big Idea 2
Drill 3 Answers and Explanations

ANSWER KEY

1. C
2. A
3. C
4. D
5. B
6. A
7. B
8. D
9. C
10. C
11. C
12. D
13. A
14. C
15. B
16. D
17. A
18. B
19. C
20. B
21. D
22. D
23. B
24. C
25. 2

ANSWERS AND EXPLANATIONS

Multiple-Choice Questions

1. **C**   A molecule is amphipathic if it has a hydrophilic region and a hydrophobic region. Answer (C) correctly identifies the hydrophilic and hydrophobic regions of a phospholipid. Answer (A) is incorrect because fatty acid tails are nonpolar, and thus, they are attracted to neither positive nor negative charges. Answer (B) is true in the sense that phospholipids are composed of carbon, hydrogen, oxygen, and phosphorous atoms, but the atomic composition does not explain why phospholipids are amphipathic. Similarly, the fatty acid tails can be either saturated or unsaturated, but that does not address the dual presence of hydrophilic and hydrophobic regions in the molecule.

2. **A**   Ester linkages are characterized by an oxygen atom bonded to an alkyl group ($-C_nH_{2n+1}$) and a carbonyl group ($-COOH$). In a triglyceride, the glycerol group is an alkyl group, and the fatty acids contain the carbonyl groups. Thus, answer (A) is correct. Peptide bonds are formed between an amino group and a carbonyl group (e.g., amino acid polymerization). There is no amino group in triglyceride, so answer (B) is incorrect. Ionic bonds are polar attractions between two oppositely charged ions. Hydrogen bonds are likewise the result of polar interactions between hydrogen and electronegative atoms. The bond between glycerol and a fatty acid is a covalent bond. Ionic bonds and hydrogen bonds are *not* covalent bonds, which eliminates answers (C) and (D).

3. **C**   Acidic solutions have low pH values and high concentrations of hydrogen ions. Answer (B) is incorrect because Solution X has a lower pH than Solution Y, and thus, Solution X more acidic than Solution Y. Because Solution X is more acidic than Solution Y, Solution X has a higher concentration of hydrogen ions. Choice (D) incorrectly identifies Solution Y as having a higher concentration of hydrogen ions. Answer (A) is a trap for test takers who focus on the fact that 4 is *twice as many* as 2. Remember, the pH scale is logarithmic rather than linear. A pH difference of 1 indicates a tenfold difference in hydrogen ions. Thus, and pH difference of 2 indicates that the difference in hydrogen ions is 100 (10×10). Answer (C) is correct.

4. **D**   Glucose is a monosaccharide, so (A) is incorrect. Sucrose is a disaccharide, and it is not a primary form of energy storage in humans. (B) is thus eliminated. Arginine is an amino acid, not a carbohydrate, so answer (C) is incorrect. Glycogen is a polymer of glucose that serves as a form of energy storage in liver and muscle cells, which makes answer (D) the correct choice.

5. **B**   Alanine is an amino acid, and all amino acids are composed of a carboxyl group, an amino group, and an R group that varies for each amino acid. A carboxyl group consists of $-COOH$, which is pictured in the upper right portion of the diagram. An amino group consists of $-NH_2$, which is pictured in the lower portion of the diagram. The R group in alanine is a methyl group, which consists of $-CH_3$. The methyl group is pictured in the left portion of the diagram. A phenyl group consists of $-C_6H_5$ and has a ring structure. There is no phenyl group in alanine, so answer (C) is correct.

6. **A** Insulin is a peptide hormone which means that it is a polypeptide. Polypeptides are synthesized by ribosomes. In eukaryotes, ribosome activity occurs either in the cytoplasm or on the surface of the rough endoplasmic reticulum. Insulin is synthesized in the rough endoplasmic reticulum rather than the cytoplasm because insulin is secreted by pancreatic cells, so insulin synthesis follows a particular pathway for secreted peptides. Therefore, answer (A) is correct. The smooth endoplasmic reticulum lacks ribosomes that characterize the rough endoplasmic reticulum, so the smooth endoplasmic reticulum is not a site of polypeptide synthesis. Lysosomes are sacs of digestive enzymes, and they are not sites of polypeptide synthesis. Answer (C) is thus incorrect. Mitochondria are organelles that synthesize ATP, not polypeptides, so answer (D) is also wrong.

7. **B** The diagram indicates that the sodium-potassium pump moves each ion from a region of lower to concentration to a region of higher concentration. Energy is required to move ions from a region of lower concentration to a region of higher concentration, and that energy is supplied by ATP hydrolysis. This is a classic example of active transport, so answer (B) is correct. Passive transport refers to the movement of solutes directly across the lipid membrane (that is, without the help of a membrane protein). Thus, answer (A) is incorrect. In facilitated transport, solutes are moved across a membrane with the help of proteins. However, in facilitated transport, the solutes move from a region of high concentration to a region of low concentration, and ATP hydrolysis is not required. That eliminates answer (C). Finally, endocytosis refers to a process in which a cell engulfs a substance by forming a pocket around the substance with the cellular membrane. The mechanism of the sodium-potassium pump does involve a membrane pocket, so answer (D) is incorrect.

8. **D** Mitochondria synthesize the primary energy storage molecule of the cell, which is adenosine triphosphate (ATP). Answer (D) is correct. Answer (A) is wrong because hydrogen peroxide (H_2O_2) is broken down in peroxisomes, not synthesized in mitochondria. Glucose is a simple sugar, and it is not synthesized in the mitochondria. Thus, answer (B) is wrong. Triglycerides are blood lipids, and they are not synthesized in the mitochondria, so answer (C) is incorrect.

9. **C** The –OH (hydroxyl) group is characteristic of alcohols, so answer (A) is true. Answer (B) is also true because cholesterol molecules are found in the membrane and serve to stabilize fluidity. The diagram should be interpreted such that each line is a bond, and the atom at the end of each bond is carbon unless another atom is indicated (O for oxygen and H for hydrogen). The total is 27 carbon atoms, so answer (D) is true. Polypeptides are short chains of amino acids which contain nitrogen atoms. Answer (C) is the correct answer because cholesterol does not contain nitrogen atoms, and cholesterol is not a polypeptide.

10. **C** *Escherichia coli* is a species of bacteria, and bacteria are prokaryotes. The primary characteristic of prokaryotes is the lack of membrane-bound organelles. Mitochondria and nuclei are membrane-bound organelles which means they are not found in prokaryotes. Thus, answers (A) and (B) are incorrect. Bacterial DNA is not located in transmembrane proteins, which are proteins that span across cellular membranes. So answer (D) is wrong. Cytosol is the liquid found inside cells, and because prokaryotes do not have nuclei, the DNA genome is located in the cytosol (C).

11. **C** The upper portion of the diagram illustrates the enzyme with substrate S bound to the active site. In the lower portion of the diagram, inhibitor I is bound to the active site, which prevents substrate S from binding. An inhibitor that binds to the active site operates through *competitive inhibition*, so answers (A) and (B) are incorrect. Because inhibitor I prevents substrate S from binding to the enzyme, the reaction rate *decreases*, which eliminates answer (D). The correct answer is (C).

12. **D** Glucose phosphorylation (answer A) is the first step of glycolysis. Glycolysis and the Krebs cycle are different stages of cellular respiration, so answer (A) is incorrect. The conversion of pyruvate to acetyl CoA is also a separate stage of cellular respiration, so answer (B) is wrong. The electron acceptor ubiquinone is part of the electron transport chain and participates in oxidative phosphorylation. Oxidative phosphorylation is not part of the Krebs cycle, so answer (C) is wrong. The first step of the Krebs cycle is the combination of acetyl coA (a two-carbon molecule) with oxalo-acetate (a four-carbon molecule) to form citrate (a six-carbon molecule). Thus, answer (D) is correct. Citrate is also referred to as citric acid, and the Krebs cycle is also referred to as the citric acid cycle.

13. **A** ATP synthase functions by joining ADP and inorganic phosphate to form ATP. That addition of a phosphate group is called phosphorylation, and the reaction is driven by protons moving down a concentration gradient with the mitochondrion. That corresponds to answer (A). Hydrolysis describes a reaction that splits ATP into ADP and inorganic phosphate, so answer (B) is wrong. ATP synthase is not involved in pumping protons, so answer (C) is eliminated. ATP synthase is not part of the electron transport chain, so answer (D) is incorrect.

14. **C** Glycolysis is summarized by the following chemical equation:

Glucose + 2 ATP + 2 NAD$^+$ → 2 Pyruvic acid + 4 ATP + 2 NADH

Two ATP molecules are consumed by the reaction, but four ATP molecules are produced. That corresponds to answer (C).

15. **B** Cellular respiration generates ATP by using oxygen as a final electron acceptor. During periods of strenuous exercise, oxygen cannot be replenished quickly enough to meet the demand for ATP energy, so an oxygen debt occurs.

16. **D** The C_3 pathway converts carbon dioxide to a three-carbon molecule called 3-phosphoglycerate. The "3" in the name of the molecule refers to the carbon attached to the phosphate group, not the total number of carbon atoms in the molecule. The Krebs cycle converts acetyl CoA to carbon di-oxide, so answer (A) is incorrect. In both the CAM pathway and the C_4 pathway, carbon dioxide is converted to oxaloacetate, so answers (B) and (C) are wrong.

17. **A** In a photosystem II reaction center, the peak absorption frequency is 680 nm, so the correspond-ing pigment is P680, not p700. Therefore, answers (C) and (D) must be eliminated. Although photosystem I can produce both ATP and NADPH, photosystem II can produce only ATP. Thus, answer (B) is wrong, and answer (A) is correct. Remember that photosystem I is *more* complex than photosystem II, and that will help you remember that photosystem II does not produce NADPH.

18. **B** The rubisco enzyme catalyzes the fixation of CO_2, but it also catalyzes the fixation of oxygen (photorespiration), which is a wasteful reaction in most plants. However, C_4 plants fixate atmospheric CO_2 using PEP carboxylase, which has a higher affinity for CO_2. C_4 plants can save energy by reducing photorespiration, so answer (A) explains the efficiency of the C_4 pathway. Because the question asks for an exception, answer (A) can be eliminated. C_4 plants utilize the C_4 and C_3 pathways, but C_4 plants can utilize the C_3 pathway more efficiently by restricting the C_3 reactions to bundle-sheath cells. Four-carbon molecules are shuttled to the bundle-sheath cells and then decarboxylation releases CO_2. This saturates the rubisco enzymes in the bundle-sheath cells with CO_2 and may increase the efficiency of carbon fixation, so answer (D) is incorrect. Also, there is very little atmospheric oxygen in the bundle-sheath cells, so photorespiration is limited. Thus, answer (D) is wrong. Answer (B) is the only choice that does not explain the efficiency of the C_4 pathway. It is true, as stated in answer (B), that rubisco catalyzes the formation of an unstable six-carbon intermediate, which quickly decays into two three-carbon molecules. However, that same process occurs in both C_4 plants and non-C_4 plants, so the process cannot explain the different efficiencies of carbon fixation in C_4 plants and non-C_4 plants.

19. **C** In plant cells, the light-dependent reactions occur in the chloroplast, which corresponds to answer (C). A vacuole is a membrane-bound organelle found in plant cells. However, the light-dependent reactions do not occur in the vacuole, so answer (A) is wrong. The cell wall is a layer that surrounds plant cells, but the light-dependent reactions occur inside the cell, so answer (B) is eliminated. Because the light-dependent reactions occur inside an organelle, they do not occur inside the cytoplasm, which eliminates answer (D).

20. **B** The light-dependent reactions occur in the thylakoids, so answer (B) is correct. The light-independent reactions (also known as the C_3 cycle) occur in the stroma, which is the fluid-filled area of a chloroplast outside the thylakoids. That eliminates answers (C) and (D). Thylakoids are only found in eukaryotic cells—usually plant cells. In eukaryotes, the Krebs cycle occurs in the mitochondria, so answer (A) is incorrect.

21. **D** There are several different types of cells that have cellular walls. However, the composition of the cell wall differs for each of the cell types. The cellular walls of plants (not bacteria) are composed of cellulose, so answer (A) is wrong. The cellular walls of fungi (not bacteria) are composed of chitin, so answer (B) is wrong. The cellular walls of diatoms (which are algae, not bacteria) are composed of silica, so answer (C) is incorrect. Bacterial cell walls are composed of peptidoglycan, so answer (D) is correct.

22. **D** Facultative anaerobes are organisms that can utilize aerobic respiration when oxygen is present or fermentation in the absence of oxygen. Because the bacteria in this question are cultured in the presence of oxygen, the cells would utilize aerobic respiration (answer D). Answer (A) is incorrect; oxygen is toxic to obligate anaerobes but not facultative anaerobes. Answer (B) is contradictory

because a cell cannot survive without cellular metabolism. Answer (C) is incorrect because fermentation is less efficient than aerobic respiration, so cells will utilize aerobic respiration if oxygen is present.

23. **B**   Facilitated transport involves the movement of a substance from an area of higher concentration to an area of lower concentration through transmembrane proteins. That support answers (B). Active transport refers to the movement of a substance from an area of lower concentration to an area of higher concentration (the reverse of what is posed in the question). Thus, answer (A) is incorrect. The sodium-potassium pump is an example of active transport because sodium and potassium are pumped from areas of lower concentration to areas of higher concentration. An antiport is a membrane protein that is involved in active transport, so answer (D) is incorrect.

24. **C**   The diagram shows that the reactants have a higher potential energy than the products. Answers (A) and (D) state that the reactants have *less* energy than the products, so those answers can be eliminated. Answer (C) is correct and (B) wrong because a reaction in which the products have less energy than the reactants is exothermic, not endothermic.

25. **2**   Glycolysis requires the input of 2 ATP molecules and produces four ATP molecules. Thus, the net production is +2. Glycolysis also produces 2 NADH molecules. The NADH molecules are converted to ATP via oxidative phosphorylation, but oxidative phosphorylation consists of a separate set of reactions from glycolysis, so those ATP molecules are not included in the answer to this question.

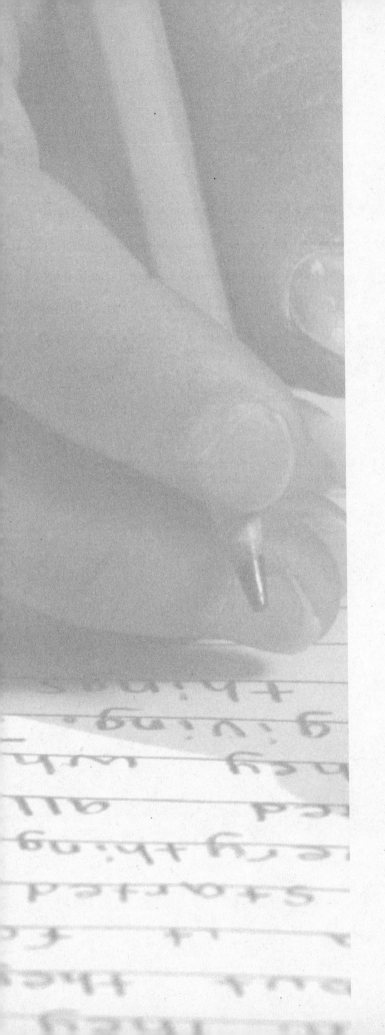

Chapter 23
Big Idea 3 Drill 3

BIG IDEA 3 DRILL 3

Multiple-Choice Questions

1. Suppose that genes X, Y, and Z are found on the same chromosome. Genes X and Y recombine with a frequency of 17%. Genes Y and Z recombine with a frequency of 9%. Genes X and Z recombine with a frequency of 8%. What is the order of these three genes on the chromosome?

   (A) X-Y-Z
   (B) Z-X-Y
   (C) X-Z-Y
   (D) Y-X-Z

2. Which of the following is a source of genetic variability?

   (A) Genetic mutation
   (B) Gene transcription
   (C) Gene expression
   (D) Natural selection

3. HIV is a retrovirus. Why is it necessary for HIV to encode for an enzyme called reverse transcriptase?

   (A) Host cells do not contain enzymes that synthesize DNA from an RNA template.
   (B) Host cells do not contain enzymes that synthesize RNA from a DNA template.
   (C) Host cells do not contain enzymes that synthesize DNA from a DNA template.
   (D) Host cells do not contain enzymes that synthesize RNA from an RNA template.

4. Which of the following sequences of DNA binds to its complementary sequence with the most hydrogen bonds?

   (A) GCATTACGC
   (B) ATTCGGCGG
   (C) ATTATAACG
   (D) GCGAGCGAC

5. Which of the following molecules carries an amino acid to the ribosome?

(A) mRNA
(B) rRNA
(C) tRNA
(D) hnRNA

6. The centrioles of a eukaryotic cell begin to form the spindle fibers during which stage of mitosis?

(A) Prophase
(B) Metaphase
(C) Anaphase
(D) Telophase

7. Gregor Mendel observed that, in pea plants, certain traits (such as yellow peas and/or tall plants) were dominant over other traits (such as green peas and/or short plants). Suppose that a tall pea plant with yellow peas is crossed with a short pea plant with green peas. Some of the offspring are short, and some of the offspring have green peas. What are the genotypes of parental plants?

(A) One plant is homozygous dominant for both traits, and the other plant is homozygous recessive for both traits.
(B) One plant is heterozygous for pea color, and the other plant is heterozygous for height.
(C) One plant is homozygous dominant for both traits, and the other plant is heterozygous for both traits.
(D) One plant is heterozygous for both traits, and the other plant is homozygous recessive for both traits.

8. HIV is a retrovirus. Which of the following is true regarding HIV?

(A) The DNA genome enters the host cell and is replicated by the host cell's machinery.
(B) The RNA genome enters the host cell and is reverse transcribed to DNA.
(C) The DNA genome enters the host cell and becomes integrated with host cell's DNA.
(D) The RNA genome enters the host cell and becomes integrated with host cell's DNA.

5'–ATTGCCGAAGCTGAC–3'

9. A single strand of DNA is symbolized above. Which of the following strands of DNA is complementary to the strand above?

(A) 5'–CAGTCGAAGCCGTTA–3'
(B) 5'–TAACGGCTTCGACTG–3'
(C) 5'–GTCAGCTTCGGCAAT–3'
(D) 5'–ATTGCCGAAGCTGAC–3'

10. Which of the following best describes DNA replication?

(A) DNA polymerase synthesizes the leading strand continuously, but the lagging strand is synthesized in fragments.
(B) DNA polymerase synthesizes the leading strand in fragments, but the lagging strand is synthesized continuously.
(C) DNA polymerase synthesizes the leading strand and lagging strand continuously.
(D) DNA polymerase synthesizes the leading strand and the lagging strand in fragments.

11. In eukaryotes, newly transcribed RNA is called heterogeneous nuclear RNA (hnRNA). Each of the following statements about hnRNA is true EXCEPT

(A) ribosomes translate hnRNA into polypeptides immediately after transcription
(B) strands of hnRNA are found in the nucleus of the cell
(C) strands of hnRNA undergo a process called splicing in which sections of RNA called introns are removed
(D) strands of hnRNA are complementary to portions of the cellular genome

12. A laboratory scientist performs the following procedure:

Step 1: Plasmids are synthesized using restriction enzymes. The plasmids contain the following segments of DNA: a promoter, a gene for erythromycin resistance, and the gene for human insulin.
Step 2: *E coli* cells are transformed with the plasmids synthesized in step 1.
Step 3: The *E. coli* cells are cultured in a nutrient-rich medium containing the antibiotic erythromycin.

What should the scientist find after step 3?

(A) There will be no *E. coli* cells remaining because *E. coli* cells are not naturally resistant to erythromycin.
(B) There will be no *E. coli* cells remaining because transformation causes cell lysis.
(C) The remaining *E. coli* cells will contain insulin mRNA and insulin protein.
(D) The remaining cells will contain insulin receptor protein and insulin protein.

13. Mutations can result from errors in the process of replication. However, some mutations cause no change in the corresponding protein sequence. How is it possible for a point mutation to occur without changing the protein sequence?

(A) It is possible to delete a single base without changing the protein sequence because some amino acids have similar physical properties.

(B) It is possible to delete a sequence of bases without changing the protein sequence because the deletion of three bases would not alter the reading frame.

(C) It is possible to substitute one base for another without changing the protein sequence because of the ambiguity of the genetic code.

(D) It is possible for to substitute one base for another without the changing the protein because of the redundancy of the genetic code.

14. If two diploid cells undergo meiosis, how many diploid daughter cells will be produced?

(A) 0
(B) 2
(C) 4
(D) 8

15. How many chromosomes does a human cell have during the anaphase stage of mitosis?

(A) 2
(B) 23
(C) 46
(D) 92

16. In which stage of meiosis does crossing over occur?

(A) Prophase I
(B) Anaphase I
(C) Prophase II
(D) Telophase II

17. The *lac* operon is a cluster of genes required for the metabolism of lactose in some bacteria. However, expression of these genes is wasteful when there is no lactose to metabolize. How do cells regulate the expression of the *lac* genes in the absence of lactose?

(A) The repressor binds to the promoter, which prevents RNA polymerase from binding to the operator.
(B) The repressor binds to the operator, which prevents RNA polymerase from binding to the promoter.
(C) The inducer cannot bind to the promoter, which prevents RNA polymerase from binding to the operator.
(D) The inducer cannot bind to the operator, which prevents RNA polymerase from binding to the promoter.

18. Which of the following processes occur during the anaphase stage of mitosis?

(A) The sister chromatids of each chromosome separate at the centromere and migrate to the opposite poles.
(B) The nucleus disappears, and the chromatin condenses into chromosomes.
(C) The nuclear membrane forms around each set of chromosomes.
(D) The chromosomes line up along the equatorial plate.

19. The snapdragon plant exhibits incomplete dominance with respect to flower color. Suppose a cross is performed between a snapdragon plant that is homozygous for red flowers (RR) and a plant that is homozygous for white flowers (rr). What color flowers will be found in the offspring?

(A) The offspring plants will have only red flowers.
(B) There will be three offspring plants with only red flowers for each plant with only white flowers.
(C) Each plant will have both red flowers and white flowers.
(D) Each plant will have pink flowers.

20. In pea plants, the tall allele (T) for the height gene is dominant, and the short allele (t) is recessive. Suppose there is a tall pea plant with unknown genotype. In order to determine the genotype, the tall plant is crossed with a short pea plant. The offspring include tall and short plants in a ratio of 3:1. What were the genotypes of parental generation?

(A) The tall plant was homozygous dominant (TT), and the short plant was homozygous recessive (tt).

(B) The tall plant was heterozygous (Tt), and the short plant was homozygous recessive (tt).

(C) The tall plant was homozygous dominant (TT), and the short plant was heterozygous (Tt).

(D) Both the tall plant and the short plant were heterozygous (Tt).

21. The genes that code for photopigments are found on the X chromosome. Missing or damaged alleles can cause color blindness. Suppose that a woman has normal vision but carries a damaged photopigment allele. If the woman marries a man with normal vision, what is the likelihood that the couple's son will be color blind?

(A) 0%
(B) 25%
(C) 50%
(D) 75%

22. Gymnosperms have a sporophyte-dominant life cycle. Which of the following is true about the sporophyte generation?

(A) The cells are haploid and produce gametophytes through mitosis.

(B) The cells are diploid and produce gametophytes through meiosis.

(C) The cells are haploid are produce gametophytes through meiosis.

(D) The cells are diploid and produce gametophytes through mitosis.

23. The pea plant has 14 chromosomes in the sporophyte generation. How many chromosomes does the plant have in the gametophyte generation?

(A) 7
(B) 14
(C) 23
(D) 28

24. Although bacteria reproduce asexually, they do exchange genetic material through several processes. Each of the following is a mechanism of genetic exchange between bacteria EXCEPT

(A) transformation
(B) conjugation
(C) transduction
(D) translation

25. Which of the following structures of a flower produces pollen grains?

(A) Anther
(B) Stigma
(C) Style
(D) Ovary

Chapter 24
Big Idea 3
Drill 3 Answers and Explanations

ANSWER KEY

1. C
2. A
3. A
4. D
5. C
6. A
7. D
8. B
9. C
10. A
11. A
12. C
13. D
14. A
15. D
16. A
17. B
18. A
19. D
20. B
21. C
22. B
23. A
24. D
25. A

ANSWERS AND EXPLANATIONS

Multiple-Choice Questions

1. **C** The frequency of recombination between two linked genes is proportional to the distance between them. So the recombination frequency is greater for genes that are far apart than it is for genes that are close together. In this question, the highest recombination frequency is between genes X and Y (17%). Thus, the distance between X and Y must be greater than the distance between either X or Y and Z. Answer (C) is the only answer choice with the maximum distance between X and Y.

2. **A** Genetic variability refers to variations from individual to individual. Genetic mutations are sources of variation from individual to individual, so answer (A) is correct. Gene transcription refers to the process of synthesizing RNA from a DNA template. In gene transcription, genes are copied to RNA, but the DNA is not altered. So answer (B) is incorrect. Gene expression refers to the transcription and translation of a gene. Like the answer to (B), these processes do not alter the DNA, so they do not create any variations in the genes. Thus, answer (C) is wrong. Finally, natural selection operates on a species by selecting certain genetic traits over others; however, natural selection is not a *source* of genetic variation. Natural selection operates because genetic variation exists. Thus, answer (D) is incorrect.

3. **A** Retroviruses are RNA viruses that go through the lysogenic life cycle. In the lysogenic cycle, a virus integrates its genetic material into the host cell's DNA. The HIV genome consists of RNA, which must be converted to DNA in order to enter the lysogenic cycle. There is no cellular mechanism in the host cell that calls for the synthesis of DNA from an RNA template, so HIV must encode for an enzyme (reverse transcriptase) to convert the viral genome to DNA. Thus, answer (A) is correct. Answers (B) and (C) are incorrect because the host cells do have the machinery to synthesize RNA from a DNA template (RNA polymerase) and the machinery to synthesize DNA from a DNA template (DNA polymerase). Answer (D) is an interesting trap answer because the host cell does not contain an enzyme that synthesizes RNA from an RNA template. However, that is not the function of reverse transcriptase, so answer (D) is true, but it does not answer this particular question.

4. **D** Complimentary DNA sequences bind together via hydrogen bonds. Adenine pairs with thymine by forming *two* hydrogen bonds, whereas cytosine pairs with guanine by forming *three* hydrogen bonds. Therefore, the number of hydrogen bonds is determined by the number of G–C bonds versus A–T bonds. The sequence in answer (A) has a combined GC count of five. Answer (B) has a GC total of six. In answer (C), the GC total is two, and in answer (D), it is seven. Thus, the sequence in answer (D) would have the most hydrogen bonds with its complementary sequence because it would have the highest number of G–C bonds.

5.   **C**   Amino acids are carried to the ribosome by tRNA (transfer RNA) molecules, so answer (C) is correct. Messenger RNA (mRNA) is synthesized from a DNA template and then translated into a peptide sequence. Therefore, answer (A) is incorrect. Ribosomal RNA (rRNA) is a component of the ribosome, so answer (B) is incorrect. Heteronuclear RNA (hnRNA) is a precursor to mRNA (prior to RNA processor). Heteronuclear RNA is not involved in the delivery of amino acids, so answer (D) is wrong.

6.   **A**   The spindle fibers begin to form during prophase, so answer (A) is correct. During metaphase, the spindle fibers are fully formed and connected to the kinetochore of each chromatid. Therefore, answer (B) is incorrect. During anaphase, the spindle fibers pull the sister chromatids apart, and the spindle fibers are gone by telophase. So answers (C) and (D) can be eliminated.

7.   **D**   The tall allele (T) is dominant over the short allele (t). And the allele for yellow peas (Y) is dominant over the allele for green peas (y). The short plant with green peas has both of the recessive traits, so that plant must be homozygous recessive. Neither answers (B) nor (C) includes a homozygous recessive plant, so those answers can be eliminated. Some of the offspring are short, and some have green peas. Thus, both of the recessive traits are represented in the offspring generation. That means that the parental plant with the dominant traits must be heterozygous, so answer (D) is correct and A incorrect. If the parental plant were homozygous dominant for both traits, then none of the offspring would have had the recessive traits.

8.   **B**   Retroviruses have RNA genomes, so answers (A) and (C) are incorrect. Answer (D) is incorrect because the RNA genome of HIV cannot be integrated with the host cell's DNA until the RNA has been reverse transcribed to DNA. Therefore, answer (B) is correct.

9.   **C**   There are several principles to remember when answering this type of question. In complimentary strands, A always pairs with T, and G always pairs with C. Complementary strands are *antiparallel*, which means the 5' end of one strand matches with the 3' end of the other strand. The DNA strand in the question and its complementary strand are depicted below.

5'–ATTGCCGAAGCTGAC–3'

3'–TAACGGCTTCGACTG–5'

The complementary strand (the lower strand) is depicted with the 3' end on the left because it is antiparallel to the upper strand. In the answer choices, however, the strands are symbolized with the 5' end on the left. The sequence of the lower strand must be reversed to match the orientation of the sequences in the answer choices. Answer (C) is the correct reversal of the lower (complementary) strand. Answer (B) is a trap answer for test takers who fail to account for the *antiparallel* structure of complementary strands.

10.   **A**   Complementary strands in double-sided DNA are antiparallel, which means the 5' end of one strand matches with the 3' end of the other strand. During replication, the complementary strands are separated, and replication begins near the point of separation. DNA polymerase only

catalyzes DNA synthesis at the 3' of a strand, so DNA synthesis always proceeds in a 5' to 3' direction. Because the template strands are antiparallel, replication can occur continuously on one of the strands (the leading strand) but not the other (the lagging strand). Instead of continuous replication, the lagging strand is synthesized in fragments called Okazaki fragments. Answer (A) is an accurate description of DNA replication. Answer (B) reverses the actual relationship between the leading strand and the lagging strand. Answer (C) incorrectly states that the lagging strand is synthesized continuously. Answer (D) incorrectly states that the leading strand is synthesized in fragments.

11. **A**   In eukaryotes, transcription occurs in the nucleus. The newly transcribed RNA must undergo processing before it is ready for translation. Thus, answer (A) is correct because hnRNA is not transcribed immediately after translation. Because transcription occurs in the nucleus, strands of hnRNA are found in the nucleus. Thus, answer (B) is wrong. RNA processing includes splicing, which is the removal of sections called introns, so answer (C) is wrong. Finally, answer (D) is wrong because strands of hnRNA are complementary to portions of the cellular genome. Strands of hnRNA are synthesized using cellular genes as templates, so the strands of hnRNA are complementary to the corresponding genes.

12. **C**   In step one, circular DNA molecules (plasmids) are synthesized. The plasmids contain two genes and a promoter. Then, the plasmids are inserted into *E. coli* cells using a process called transformation. Transformation allows DNA to enter the cells, but it does not cause cell lysis, so answer (B) is incorrect. In step 3, the *E. coli* cells are cultured in a medium containing erythromycin. The plasmids contained a gene conferring erythromycin resistance, and the promoter ensures that the gene is being expressed. So any cell that took up the plasmid DNA acquired resistance to erythromycin and will survive in the culture medium. That eliminates answer (A). The antibiotic will kill the cells that failed to take up a plasmid, and the remaining cells will contain the plasmid with both genes. The cells with the plasmid will also express the insulin gene, which means that the cells will transcribe the gene into mRNA and translate the mRNA into a protein. That corresponds to answer (C). Answer (D) is incorrect because it states that the cells will contain insulin receptor protein. Insulin receptor is a different human protein, and the gene for insulin receptor was not included in the plasmids. So the cells would not be expected to contain a human protein without the introduction of that gene through a plasmid.

13. **D**   A point mutation is the substitution of one base for another. Answers (A) and (B) are incorrect because they refer to deletions, which are distinct from point mutations. The genetic code is redundant but not ambiguous, so answer (C) is wrong and answer (D) is correct. The genetic code is not ambiguous because each codon on corresponds with exactly one amino acid. A particular codon will always code for the same amino acid. However, there may be several different amino acids that code for the same amino acid. That characteristic of the genetic code is called redundancy (i.e., there are redundant codes for some amino acids).

14. **A** Meiosis is a process in which a single diploid cell undergoes two rounds of cell division to produce four haploid daughter cells. This question asks how many *diploid* daughter cells would be produced if two diploid cells underwent meiosis. The process of meiosis produces only haploid cells, so the answer is (A): Zero *diploid* cells would be produced.

15. **D** Normally, a human cell has 46 chromosomes. Prior to mitosis, the cell replicates its genome, so each chromosome consists of two sister chromatids. During anaphase, the sister chromatids separate toward opposite ends of the cell. At the point of separation, the sister chromatids because separate chromosomes, so the total number of chromosomes doubles to 92. That corresponds to answer (D). Half of those chromosomes will be enclosed in one daughter cell and half in the other daughter cell, so the number of chromosomes in each cell will return to 46 after cytokinesis.

16. **A** Crossing over is the exchange of DNA between homologous chromosomes. This occurs during prophase I after the homologous chromosomes line up side by side. Thus, answer (A) is correct. During anaphase I, the homologous chromosomes separate, so crossing over can no longer occur. Thus, answer (B) is incorrect. Once the homologous chromosomes separate, they do not reconnect during meiosis, so answers (C) and (D) are also incorrect.

17. **B** The *lac* genes are regulated by a repressor. In the absence of lactose, the repressor binds to the operon. Answer (A) is incorrect because the repressor does not bind to the promoter. The promoter is the area of DNA where the RNA polymerase binds, but the binding of RNA polymerase is inhibited by the repressor. The inducer of the *lac* operon is lactose. When lactose is present, it binds to the repressor, which inhibits the repressor from binding to the operon. Answers (C) and (D) are incorrect because the question stipulated that the inducer (lactose) was not present. Plus, the inducer of the *lac* operon binds to the repressor, not to the promoter or the operator.

18. **A** Anaphase is the stage in which the sister chromatids of each chromosome separate. That corresponds to answer (A). Answer (B) describes prophase. Answer (C) describes telophase, and answer (D) describes metaphase.

19. **D** Incomplete dominance is an exception to the rules of Mendelian genetics. Incomplete dominance occurs when the heterozygote phenotype is a blend of the phenotypes of the homozygotes. In this question, all of the offspring will be heterozygous (Rr). Even though the red gene is designated with a capital R, the question stated that this is an example of incomplete dominance. So, the offspring will have neither red nor white flowers. Instead, the heterozygotes will have flowers with a blended color: pink. That corresponds to answer (D). Answer (A) is a trap that plays on the traditional Mendelian genetics. The offspring would have only red flowers if this was an example of dominance, but because incomplete dominance is an exception to the rules of dominance, answer (A) is incorrect. Answer (B) is referencing the 3:1 ratio expected when crossing heterozygotes. In this case, the Mendelian ratios do not apply, and the parental plants are not heterozygotes. So answer (B) is wrong. Answer (C) describes a phenomenon called co-dominance. Co-dominance is a different exception to Mendelian genetics, so answer (C) is eliminated.

20. **B** Each plant has two alleles for the height gene. There are three possible combinations of alleles: TT, Tt, and tt. Because the tall gene is dominant, the homozygous dominant (TT) and heterozygous (Tt) plants are tall. The only combination of genes resulting in a short plant is homozygous recessive (tt). So the short plant in the parental generation must have been homozygous recessive. That eliminates answers (C) and (D). If the tall plant was homozygous dominant (TT), then all of the offspring would be heterozygous (Tt). That would result in only tall offspring, so answer (A) is wrong. If the tall plant is heterozygous (Tt), then some of the offspring will be heterozygous (Tt), and some of the offspring will be homozygous (tt). That corresponds with answer (B).

21. **C** The genes for photopigments are sex-linked because they are found on the sex chromosome. The question stipulates that the couple will have a son, so his genotype must be XY. The father contributes the Y chromosome, and the mother contributes the X chromosome. The mother is a carrier of the damaged photopigment allele, but she has normal vision, so one of her X chromosomes has the normal allele, and the other has the damaged allele. There is a 50% chance that the mother will donate the damaged allele. Because the son does not have another X chromosome to mask the effects of a damaged gene, there is a 50% chance that the son will be colorblind.

22. **B** The sporophyte generation is diploid, so answers (A) and (C) are incorrect. The gametophyte generation is haploid. The diploid sporophyte cells undergo meiosis in order to form haploid cells, so answer (B) is correct. Answer (D) is wrong because the mitotic division of a diploid cell would produce another diploid cell.

23. **A** The sporophyte generation is the diploid generation of the plant life cycle. The plant has 2n chromosomes during the sporophyte generation. The gametophyte generation is the haploid generation (1n), so the gametophyte will have half the chromosomes of the diploid generation. Half of 14 is 7, so answer (A) is correct. Answer (C) is the number of chromosomes in a humane gamete.

24. **D** Transformation occurs when a bacterium takes in naked DNA; this is a mechanism of genetic exchange, so answer (A) can be eliminated. Conjugation is a process in which bacteria share genetic information via a bridge called the pilus. Conjugation is also a mechanism of genetic exchange, so answer (B) is wrong. Transduction is the exchange of genetic information between bacteria via a virus. That is also a mechanism of genetic exchange, so answer (C) is wrong. Translation is the process of synthesizing polypeptides from mRNA. Translation does not involve the exchange of genetic material between cells, so answer (D) is correct.

25. **A** The anther is the structure that produces pollen grains, so answer (A) is correct. The stigma is the sticky portion of the pistil that captures pollen grains, so answer (B) is wrong. The style connects the stigma to the ovary, but the style does not produce pollen. So answer (C) is wrong. The ovary is where fertilization occurs, but it is not the structure that produces pollen grains. Thus, answer (D) is wrong.

Chapter 25
Big Idea 4 Drill 3

BIG IDEA 4 DRILL 3

Multiple-Choice Questions

1. The function of salivary amylase is to break down

   (A) proteins into smaller peptides
   (B) lipids into fatty acids and glycerol
   (C) nucleic acids into nucleotides
   (D) starch into maltose

2. Which of the following is a pancreatic enzyme in humans that breaks down proteins?

   (A) Pancreatic lipase
   (B) Pancreatic amylase
   (C) Chymotrypsin
   (D) Pepsin

3. What is the function of gastrin in humans?

   (A) Gastrin is an enzyme that breaks down proteins.
   (B) Gastrin is an enzyme that breaks down carbohydrates.
   (C) Gastrin is a hormone that stimulates stomach cells.
   (D) Gastrin is a hormone that stimulates pancreatic cells.

4. What is the mechanism of gas exchange in human lungs?

   (A) Primary active transport
   (B) Secondary active transport
   (C) Passive diffusion
   (D) Facilitated diffusion

5. In the capillaries of human muscles, hemoglobin

    (A) delivers CO_2 and binds oxygen
    (B) delivers oxygen and binds CO_2
    (C) delivers CO_2 and oxygen
    (D) delivers iron and binds oxygen

6. Gas exchange occurs in which of the following anatomical structures?

    (A) Larynx
    (B) Trachea
    (C) Bronchus
    (D) Alveoli

7. Which of the following is true regarding pulmonary circulation in humans?

    (A) Oxygenated blood travels to the lungs, and deoxygenated blood returns to the heart.
    (B) Deoxygenated blood travels to the lungs, and oxygenated blood returns to the heart.
    (C) Oxygenated blood travels to tissues throughout the body, and deoxygenated blood returns to the heart.
    (D) Deoxygenated blood travels to tissues throughout the body, and oxygenated blood returns to the heart.

8. There are four human blood groups: A, B, AB, and O. Individuals with the AB blood type have blood cells with both the A and B surface antigens. This is an example of

    (A) codominance
    (B) incomplete dominance
    (C) pleiotropy
    (D) epistasis

9. Which tissue initiates the electrical impulses that trigger cardiac contraction in humans?

(A) Atrioventricular node
(B) Sinoatrial node
(C) Purkinje fibers
(D) Bundle of His

10. In a human immune system, cytotoxic T cells

(A) activate B lymphocytes
(B) are responsible for long-term immunity
(C) recognize and kill infected cells
(D) engulf and digest pathogens

11. Antibodies are produced by

(A) T lymphocytes
(B) B lymphocytes
(C) cytotoxic T cells
(D) macrophages

12. One function of the lymphatic system is to

(A) deliver oxygenated blood to body tissues and deoxygenated blood to the lungs
(B) filter nitrogenous waste from the blood
(C) remove excess fluid from body tissue
(D) detoxify the blood

13. In humans, nitrogenous wastes are excreted as

    (A) ammonia
    (B) urea
    (C) uric acid
    (D) methylamine

14. Which of the following occurs in the proximal convoluted tubule of a human kidney?

    (A) Blood is filtered.
    (B) Ions are secreted into the filtrate.
    (C) Water, nutrients, and salts and reabsorbed into the blood.
    (D) Water, nutrients, and salts are secreted into the filtrate.

15. After leaving a human kidney, urine travels through each of the following structures EXCEPT

    (A) ureter
    (B) bladder
    (C) urethra
    (D) efferent arteriole

16. An action potential of a neuron initiates in the

    (A) dendrite
    (B) axon
    (C) axon bulb
    (D) nucleus

17. During an action potential, depolarization of a neuron is caused by the

   (A) opening of potassium channels
   (B) closing of potassium channels
   (C) opening of voltage-gated sodium channels
   (D) closing of voltage-gated sodium channels

18. Each of the following is a neurotransmitter EXCEPT

   (A) GABA
   (B) acetylcholine
   (C) norepinephrine
   (D) acetylcholinesterase

19. Which of the following proteins plays a role in the contraction of skeletal muscles?

   (A) Calcium
   (B) Troponin
   (C) ATP
   (D) Acetylcholine

20. Which of the following types of tissue attaches bones to muscles?

   (A) Tendon
   (B) Ligament
   (C) Adipose
   (D) Collagen

21. In humans, osteoblasts are responsible for

   (A) bone formation
   (B) bone resorption
   (C) phagocytosis of pathogens
   (D) formation of the myelin sheath

22. Each of the following hormones is secreted from the anterior pituitary gland EXCEPT

   (A) adrenocorticotropic hormone
   (B) follicle-stimulating hormone
   (C) luteinizing hormone
   (D) oxytocin

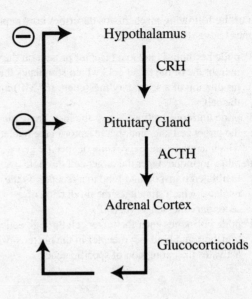

23. The diagram above symbolizes some of the relationships between the hypothalamus, pituitary gland, and adrenal gland. Which of the following terms best describes the mechanism depicted above?

   (A) Negative feedback
   (B) Positive feedback
   (C) Allosteric inhibition
   (D) Competitive inhibition

24. Which of the following mechanisms describes how peptide hormones trigger their target cells?

   (A) Peptide hormones bind to a receptor protein on the cell membrane of the target cell, which stimulates the production of a secondary messenger (cAMP) inside the cell.

   (B) Peptide hormones diffuse across the membrane of the target cell and bind to a receptor in the nucleus, which activates the expression of specific genes.

   (C) Peptide hormones enter the target cell through facilitated transport and bind to a receptor in the nucleus, which stimulates the production of a secondary messenger (cAMP).

   (D) Peptide hormones enter the target cell through active transport and bind to a receptor in the nucleus, which activates the expression of specific genes.

25. The luteal surge of the female menstrual cycle triggers

   (A) the follicle to grow
   (B) the endometrium to grow
   (C) the endometrium to thicken
   (D) the follicle to burst, releasing the ovum

26. Which of the following hormones is secreted during pregnancy to help maintain the uterine lining?

   (A) Follicle-stimulating hormone
   (B) Luteinizing hormone
   (C) Human chorionic gonadotropin
   (D) Vasopressin

27. Gastrulation is a stage of human embryonic development in which the cells of the embryo differentiate to form germ layers. Each of the following is a germ layer of the gastrula EXCEPT the

(A) ectoderm
(B) mesoderm
(C) blastoderm
(D) endoderm

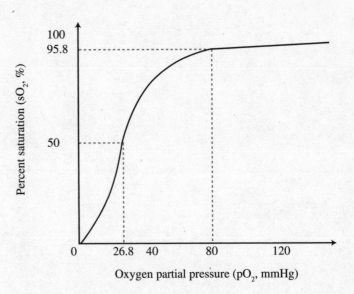

28. The protein hemoglobin binds oxygen to transport it around the body. The diagram above compares the saturation of hemoglobin with the pressure of oxygen. Which of the following describes the function of hemoglobin in the capillaries of the lungs?

(A) The partial pressure of oxygen is relatively high in the lungs, and that causes the hemoglobin proteins to become less saturated.
(B) The partial pressure of oxygen is relatively high in the lungs, and that causes the hemoglobin proteins to become more saturated.
(C) The partial pressure of oxygen is relatively low in the lungs, and that causes the hemoglobin proteins to become less saturated.
(D) The partial pressure of oxygen is relatively low in the lungs, and that causes the hemoglobin to become more saturated.

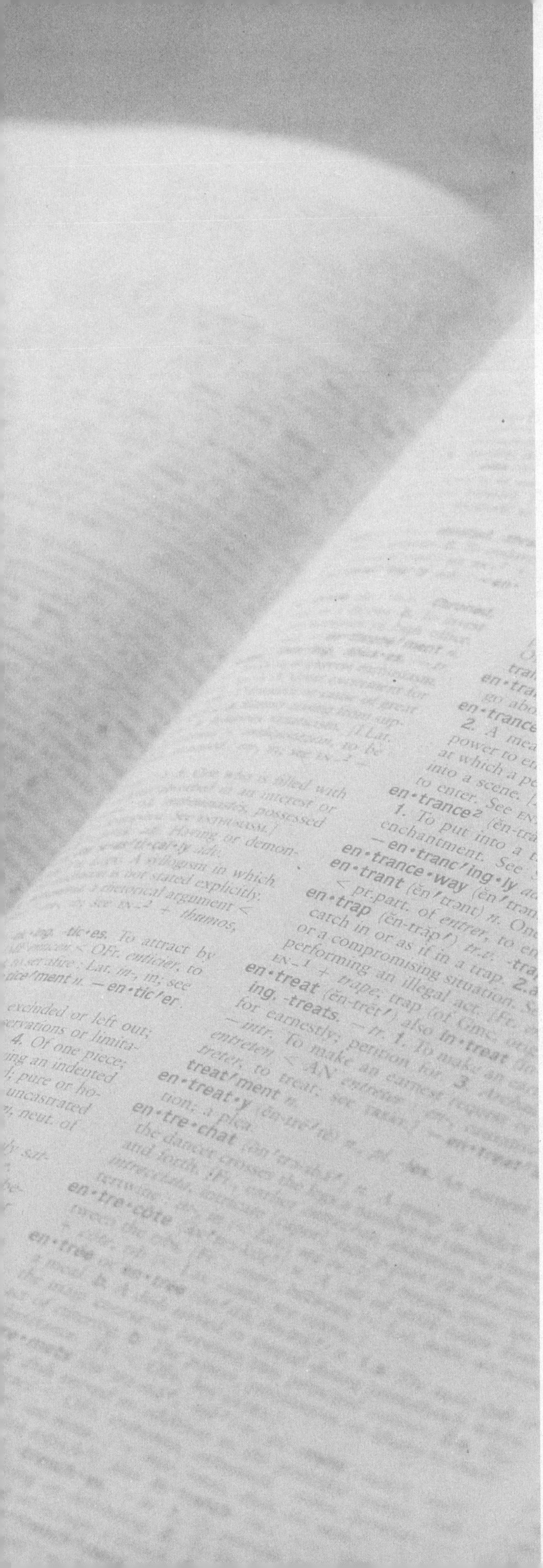

Chapter 26
Big Idea 4
Drill 3 Answers and Explanations

ANSWER KEY

1. D
2. C
3. C
4. C
5. B
6. D
7. B
8. A
9. B
10. C
11. B
12. C
13. B
14. C
15. D
16. A
17. C
18. D
19. B
20. A
21. A
22. D
23. A
24. A
25. D
26. C
27. C
28. B

ANSWERS AND EXPLANATIONS

Multiple-Choice Questions

1. **D** Salivary amylase is an enzyme found in saliva that breaks down starch into maltose. Thus, answer (D) is correct. Answer (A) is wrong because it describes the function of pepsin. Answer (B) is incorrect because it describes the function of pancreatic lipase. Answer (C) is incorrect because it describes the function of ribonuclease and deoxyribonuclease.

2. **C** Pancreatic lipase is a pancreatic enzyme, but it breaks down lipids into fatty acids and glycerol. So answer (A) is wrong. Pancreatic amylase breaks down starch into disaccharides, so answer (B) is wrong. Pepsin is an enzyme that breaks down proteins; however, pepsin is secreted by the stomach, not the pancreas, so answer (D) is wrong. Chymotrypsin is a pancreatic enzyme that breaks down proteins into dipeptides, so answer (C) is correct.

3. **C** Gastrin is a hormone that plays a role in human digestion. Gastrin is not an enzyme, so answers (A) and (B) can be eliminated. The function of gastrin is to stimulate stomach cells to produce gastric juice. Therefore, answer (C) is correct, and (D) is wrong.

4. **C** Through cellular metabolism, human cells consume oxygen and produce carbon dioxide. This creates a concentration gradient for both oxygen and carbon dioxide in the lung cells. Carbon dioxide and oxygen are small, non-polar molecules, so they can both diffuse across the cellular membranes of the lung cells. Because of the concentration gradient in the lungs, gas exchange occurs via passive diffusion, so answer (C) is correct. Active transport requires the expenditure of cellular energy, which is not required in gas exchange because the concentration gradient drives the movement of gases in the lungs. Therefore, answers (A) and (B) are incorrect. Facilitated transport is the movement of molecules across a membrane through transmembrane proteins. Transmembrane proteins are not necessary for gas exchange in the lungs because oxygen and carbon dioxide are small, non-polar molecules that can diffuse directly across the membrane. Therefore, answer (D) is wrong.

5. **B** Hemoglobin is a protein that aids in the transportation of gases in the circulatory system. In the capillaries of the lungs, hemoglobin binds oxygen to transport it around the body. In the muscle cells, however, hemoglobin delivers oxygen to the muscle tissues that need energy. In addition to delivering oxygen, hemoglobin binds CO_2 to transport it away from the muscles to expel from the lungs. Therefore, answer (B) is correct. Answer (A) describes the role of hemoglobin in the lungs, not muscles, so answer (A) is wrong. Hemoglobin does not deliver CO_2 to muscle cells, so answer (C) is wrong. Although hemoglobin contains iron, hemoglobin does not deliver the iron, so answer (D) is incorrect.

6. **D** The conducting zone of the respiratory system includes the larynx, trachea, and bronchi. Gas exchange does not occur in the conducting zone, so answers (A), (B), and (C) are incorrect. The alveoli are part of the respiratory zone, which is where gas exchange occurs.

7. **B** There are two circulations in the cardiovascular system. In systemic circulation, blood travels from the heart to tissues throughout the body. In pulmonary circulation, blood travels from the heart to the lungs. Therefore, answers (C) and (D) can be eliminated. Deoxygenated blood flows from the heart to the lungs in order to facilitate gas exchange. The blood becomes oxygenated in the lungs, and oxygenated blood returns to the heart. Therefore, answer (B) is correct, and (A) is wrong.

8. **A** Codominance is a phenomenon in which heterozygotes express the phenotypes associated with both alleles. Individuals who are heterozygous (AB) express both the A surface antigen and the B surface antigen. This is an exception to Mendelian genetics, and it is an example of codominance. Thus, answer (A) is correct. Incomplete dominance occurs when the heterozygous phenotype is a distinct blend of the phenotypes associated with both alleles. Blood cells do not express a blended AB antigen; rather, blood cells express *both* the A and B antigens. Thus, answer (B) is wrong. Pleiotropy occurs when one gene affects multiple phenotypic trains. The scope of this question is limited to one phenotypic train (blood type), so answer (C) is wrong. Epistasis occurs when the expression of one gene depends on the expression of another gene. The A and B surface antigens are coded by different alleles of the same gene, so this is not an example of one gene affecting another gene. Therefore, answer (D) is incorrect.

9. **B** The electrical impulse of the heart is generated in the sinoatrial node, so answer (B) is correct. From the sinoatrial node, the impulse spreads through both and directly to the atrioventricular node. From there, the impulse spreads to the bundle of His and then to the Purkinje fibers. Thus, answers (A), (C), and (D) are incorrect.

10. **C** The role of cytotoxic T cells is to recognize and kill infected cells. Therefore, answer (C) is correct. Answer (A) is incorrect because it describes *helper* T cells, which activate B lymphocytes. Answer (B) is incorrect because it describes *memory* T cells and memory B cells. Answer (D) is wrong because it describes phagocytes such as macrophages.

11. **B** Antibodies are produced by B lymphocytes, so answer (B) is correct. Answer (A) is wrong because T lymphocytes fight infection and activate B lymphocytes, but T lymphocytes do not produce antibodies. The cytotoxic T cell is a type of T lymphocyte, so answer (C) is wrong for the same reason answer (A) is wrong. Macrophages are phagocytes that do not produce antibodies, so answer (D) is wrong.

12. **C** The lymphatic system refers to the conduits that carry lymph (excess fluid) from the tissues of the body to the heart. Thus, answer (C) is correct. Answer (A) is wrong because it describes the function of the cardiovascular system. Answer (B) is incorrect because it describes the renal system. Answer (D) is wrong because it describes the function of the liver.

13. **B** Protein degradation creates nitrogenous wastes that individuals must eliminate. Ammonia is toxic, however, so individuals do not store or eliminate nitrogen as ammonia. Thus, answer (A) is wrong. Instead, mammals convert nitrogenous wastes to urea, which is practically non-toxic. Thus, answer (B) is correct. Some birds and reptiles convert nitrogenous waste to uric acid, but this pathway is not found in mammals, so answer (C) is wrong. Methylamine is a nitrogenous compound, but it is not part of the mammalian excretory pathway. Thus, answer (D) is incorrect.

14. **C** In the human kidney, blood is filtered by the glomerulus, so answer (A) is incorrect. Ions are secreted into the filtrate in the distal convoluted tubule, so answer (B) is incorrect. Water, nutrients, and salts are substances that the human body conserves, so the function of the proximal convoluted tubule is to *reabsorb* those substances. Thus, answer (C) is correct, and answer (D) is wrong.

15. **D** After leaving the kidney, urine travels through the ureters to the bladder. Urine is then stored in the bladder until it is eliminated through the urethra. Therefore, answers (A), (B), and (C) are incorrect. Urine never travels efferent arteriole, so answer (D) is correct. Blood flows away from the glomerulus through the efferent arteriole.

16. **A** The action potential of a neuron is generated in the dendrite, so answer (A) is correct. From there, the action potential spreads through the cell body and down the axon. Therefore, answer (B) is incorrect. The axon bulb is at the end of the axon, so answer (C) is wrong. Actions potentials are waves of depolarization across the membrane of a cell. The nucleus is an internal cellular organelle that protects the genome. Thus, answer (D) is wrong.

17. **C** Because the resting membrane potential of a neuron is negative, the cell's charge must increase to depolarize. This is achieved by opening voltage-gated sodium channels, which allows Na^+ ions to enter the cell. Therefore, answer (C) is correct. Answer (A) describes the mechanism of repolarization, in which the charge of the cell decreases due to an outflow of potassium channels. So answer (A) is incorrect. Closing the aforementioned channels signals the end of a stage of the action potential, not a mechanism for the next stage, so answers (B) and (D) are incorrect.

18. **D** GABA, acetylcholine, and norepinephrine are all examples of neurotransmitters, so answers (A), (B), and (C) are wrong. Acetylcholinesterase is not a neurotransmitter, so answer (D) is correct. Acetylcholinesterase is an enzyme that breaks down acetylcholine in the synaptic cleft. The suffix -ase is often used in biochemistry to form the names of enzymes.

19. **B** This is a tricky question because all of the answers are molecules that play a role in the contraction of skeletal muscles. However, only one of the answers is a *protein*. Calcium ions bind to troponin, exposing the myosin-binding sites on the actin filaments. Calcium ions are not proteins, so answer (A) is incorrect. Muscle contraction entails the hydrolysis of ATP, but ATP is a nucleotide, not a protein, so answer (C) is wrong. Acetylcholine is a neurotransmitter that initiates an action potential in skeletal muscle cells. Acetylcholine is an ester, not a protein, so answer (D) is wrong. Troponin is a protein that binds with calcium, which exposes the myosin binding sites. Thus, answer (B) is correct.

20. **A** A tendon is a type of connective tissue that usually connects muscle to bone, so answer (A) is correct. A ligament is a tendon that connects a bone to another bone, so answer (B) is wrong. Adipose tissue is body fat, so answer (C) is incorrect. Collagen is a protein found in connective tissue, but collagen is not a type of tissue, so answer (D) can be eliminated.

21. **A** Osteoblasts are responsible for bone formation, so answer (A) is correct. Bone resorption is carried out by osteoclasts, so answer (B) is incorrect. Answer (C) is wrong because it describes the function of macrophages. Myelinating Schwann cells are responsible for formation of the myelin sheath, so answer (D) is wrong.

22. **D** The anterior pituitary gland secretes adrenocorticotropic hormone, follicle-stimulating hormone, and luteinizing hormone, so answers (A), (B), and (C) are incorrect. Oxytocin, however, is secreted from the *posterior* pituitary, which is distinct from the *anterior* pituitary. Therefore, answer (D) is correct.

23. **A** Hormones frequently operate by a negative feedback system in which excess amounts of a hormone will signal an endocrine gland to shut down production. Here, the hypothalamus secretes CRH, which stimulates the pituitary gland to secrete ACTH, which in turn stimulates the adrenal cortex to secrete glucocorticoids. Then, the increased level of glucocorticoids feeds back to shut down production at the hypothalamus and pituitary gland. Thus, answer (A) is correct. In positive feedback, the hormone causes an *increase* in secretion from the endocrine glands, whereas here there is a decrease, so answer (B) is incorrect. Allosteric inhibition and competitive inhibition refer to the regulation of enzymes, not hormones, so answers (C) and (D) are incorrect.

24. **A** Peptide hormones are large, polar molecules, so they are unable to diffuse across the cellular membrane. Thus, answer (B) is incorrect. Peptide hormones function by binding to receptors on the cell membrane of the target cell which stimulates the production of cAMP inside the cell. That corresponds with the correct answer, (A). Peptide hormones do not function by entering the target cells, so answers (C) and (D) are incorrect.

25. **D** One characteristic of the female menstrual cycle is a surge of luteinizing hormone secreted by the pituitary gland. This luteal surge causes the follicle to burst, releasing the ovum (ovulation). Thus, answer (D) is correct. Follicular growth is stimulated by FSH, not LH, so answer (A) is wrong. Endometrial growth is stimulated by estrogen, not LH, so answer (B) is wrong. Finally, answer (C) is incorrect because the thickening of the endometrium is stimulated by progesterone, not LH.

26. **C** When pregnancy occurs, the extraembryonic tissue of the fetus secretes human chorionic gonadotropin which helps maintain the uterine lining. That corresponds to answer (C). Follicle-stimulating hormone stimulates follicular growth, so answer (A) is incorrect. Luteinizing hormone stimulates ovulation, so answer (B) can be eliminated. Vasopressin regulates water retention and blood vessel constriction, so answer (D) is wrong.

27. **C** The three germ layers of the gastrula are: ectoderm, mesoderm, and endoderm. Therefore, answers (A), (B), and (D) are incorrect. A blastoderm is a layer of cells found in yolky bird eggs, not in human embryonic gastrula, so answer (C) is correct.

28. **B** The partial pressure of oxygen is higher in the lungs than in it is in other tissue because tissue cells consume oxygen. Lung cells are exposed to the oxygen in the air, so the partial pressure of oxygen in the lungs is relatively high. Thus, answers (C) and (D) are incorrect. The diagram above shows that the saturation of hemoglobin increases with as the partial pressure of oxygen increases. Therefore, answer (B) is correct. Answer (C) incorrect states that hemoglobin proteins become *less* saturated in the lungs.

Chapter 27
Big Idea 1 Drill 4

BIG IDEA 1 DRILL 4

Multiple-Choice Questions

1. In the classic example of evolution of the peppered moth to the Industrial Revolution. How would the results of this study have been different, if factories produced white or light gray ash and soot rather than black?

    (A) There would be no change to the results of the experiment.
    (B) There would been added selection pressure for more white-bodied spotted moths and against black-bodied spotted moths.
    (C) There would been added selection pressure for more black-bodied spotted moths and against white-bodied spotted moths.
    (D) There would have been an increase in the frequency of both black-bodied and white-bodied spotted moths.

2. The RNA world hypothesis suggests that current life which is DNA-based originally evolved from RNA-based organisms. Which of the following would best support this view?

    (A) The genetics of bacteria
    (B) The genetics of viruses
    (C) The genetics of archaea
    (D) The genetics of eukaryotes such as yeast

3. An archaeological fin reveals the bones of a giant sloth. A scientist uses carbon-14 dating to predict their age. The experiment revealed that carbon-14 had undergone 3 half-lives in the bones. If the half-life of carbon-14 is approximately 5,800 years, approximately how old are the bones?

    (A) 193 years
    (B) 1,933 years
    (C) 5,800 years
    (D) 17,400 years

4. The unique chemical nature of water is critical for the survival of aquatic life in more temperature environments. All of the following characteristics of water are important for the survival of aquatic life EXCEPT

    (A) water is a strong solvent due to its polar nature which allows for solubility of salts and gases.
    (B) water has a high specific heat which prevents rapid changes in temperature.
    (C) water is denser as a liquid than a solid which ensures that ice forms at the surface rather than the bottom.
    (D) water has low surface tension making it easier for aquatic life to walk on its surface.

The increase in greenhouse gases in the atmosphere has contributed to changes in the global climate and likely weather patterns over the last 150 years. The graph below displays the carbon dioxide (CO_2) concentration and average temperature since 1880.

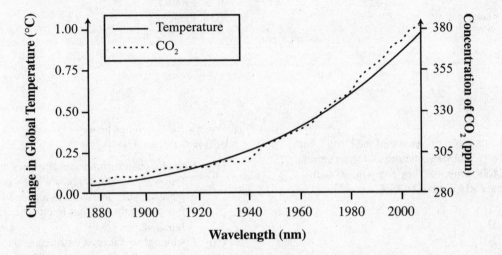

5. Based on the figure above, what is the relationship between atmospheric carbon dioxide concentrations and changes in temperature between 1920 and today?

   (A) As atmospheric carbon dioxide concentrations increased, the temperatures decreased.
   (B) As atmospheric carbon dioxide concentrations increased, the temperatures increased.
   (C) Both atmospheric carbon dioxide and temperatures declined between 1920 and today.
   (D) The relationship cannot be determined from the data provided.

6. Which of the following best explains how carbon dioxide impacts global climate?

   (A) Carbon dioxide reacts with ozone in the atmosphere creating holes in the ozone layer allowing more ionizing radiation to enter the atmosphere, which heats the planet.
   (B) Carbon dioxide gas in the atmosphere limits the amount of UV light from the sun reaching the surface, which cools the planet.
   (C) Carbon dioxide gas absorbs the light reflected from the surface preventing it from being released back into space, which heats the planet.
   (D) Carbon dioxide reacts with oxygen in the atmosphere in an endothermic reaction forming carbonate salts, which cools the planet.

7. By approximately what percent has the atmospheric carbon dioxide concentrations increased between 1900 and 2000?

   (A) 10%
   (B) 20%
   (C) 33%
   (D) 50%

8. The carbon cycle is a critical component of the maintenance of the biosphere. All of the following organisms use carbon dioxide as a substrate and produce oxygen is waste product EXCEPT:

   (A) cyanobacteria
   (B) algae
   (C) flowering plants
   (D) amoebae

9. Scientists believe that the atmosphere of the Earth went through several different stages before taking its current form. All of the following were key components of the atmosphere of the early Earth EXCEPT:

   (A) methane (CH_4)
   (B) hydrogen (H_2)
   (C) ammonia (NH_3)
   (D) oxygen (O_2)

10. The cell wall provides structure and stability to the architecture of prokaryotic and eukaryotic cells. Which of the following organisms does not normally have a cell wall?

    (A) Bacteria
    (B) Fungi
    (C) Protists
    (D) Plants

11. What is the difference between homologous structures and analogous structures?

    (A) Homologous structures are structures that share functional similarity and are likely due to a common ancestor, whereas analogous structures are independently evolved structures toward a common function.
    (B) Analogous structures are structures that share functional similarity and are likely due to a common ancestor, whereas homologous structures are independently evolved structures toward a common function.
    (C) Both homologous structures and analogous structures share functional similarity due to a common ancestor, though analogous structures result in new uses of the evolved structures.
    (D) There is no difference; they are the same thing.

12. An abrupt change in the environment selects for organisms that have traits on the extremes regarding size and against those of average height. Which type of selection is described?

(A) Directional selection
(B) Stabilizing selection
(C) Disruptive selection
(D) Sexual selection

13. The classic behavior example of baby chicks associating the first moving object they see as their mother is an example what type of learning?

(A) Instinct
(B) Classical conditioning
(C) Operant conditioning
(D) Imprinting

14. Tropism is a turning response to a stimulus. Which of the following is response by plants to touch or physical changes in environment?

(A) Phototropism
(B) Gravitropism
(C) Thigmotropism
(D) Chemotropism

15. Many algae are photoautotrophs. Which of the following is true regarding this group of algae?

(A) These algae produce complex organic compounds from simple substances and obtain energy by oxidation.
(B) These algae produce complex organic compounds from simple substances and obtain energy by photon capture.
(C) These algae metabolize complex organic compounds and obtain energy by oxidation.
(D) These algae metabolize complex organic compounds and obtain energy by photon capture.

16. The movement of a plant in response to an environmental stimulus is a tropism. Climbing plants, such as vines, can coil around supporting objects like walls or trellises by moving or growing in response to contact with the supporting objects. This movement is an example of which of the following?

(A) Phototropism
(B) Thigmotropism
(C) Gravitropism
(D) Thermotropism

17. Which of the following is an example of habituation?

(A) Pavlov's dogs began to salivate upon hearing a bell ring after repeated pairings of the bell ringing and food.
(B) Goslings follow the first the first moving object they see during a critical period after hatching.
(C) A colony of prairie dogs gives fewer and fewer alarm calls after repeated exposure to humans.
(D) A dog sits on command after being rewarded with treats in the past.

18. Each of the following describes a social behavior of animals EXCEPT

(A) circadian rhythm
(B) dominance hierarchies
(C) territoriality
(D) altruistic behavior

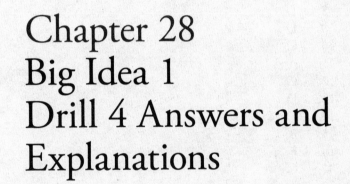

Chapter 28
Big Idea 1
Drill 4 Answers and Explanations

ANSWER KEY

1. B
2. B
3. D
4. D
5. A
6. C
7. C
8. D
9. D
10. C
11. A
12. C
13. D
14. C
15. B
16. B
17. C
18. A

ANSWERS AND EXPLANATIONS

Multiple-Choice Questions

1. **B** If the color of ash or soot produced by the Industrial Revolution were white or light gray, this would likely reverse the trend observe applying additional selection against the black moths.

2. **B** Only viruses exhibit genomes today that are RNA-based. Although all of life uses RNA, the mechanism by which an RNA-based organism would function would likely be closest to an RNA virus.

3. **D** The age of the bones will be equal to 3 times the half-life (5,800 years) or 17,400 years.

4. **D** Water has high surface tension due to its high intermolecular forces. This is why some insects are able to walk on the surface without breaking it. Water is a strong solvent (A), has a high specific heat (B), and water is more dense (C).

5. **A** Based on the figure, as the carbon dioxide concentration increased between 1920 and the present day, the global temperatures have increased.

6. **C** Carbon dioxide is a greenhouse gas which traps the light from the sun in the atmosphere. As light is reflected off of the surface it is trapped by carbon dioxide gas in the atmosphere. This leads to increases in temperature.

7. **C** The concentration of carbon dioxide in the atmosphere increased from approximately 285 ppm in 1900 to 375 ppm in 2000. The concentration therefore increased approximately 33% (~300 to 400).

8. **D** Protists such as amoebae are not photosynthetic and therefore would not use carbon dioxide as a substrate. Cyanobacteria (A), algae (B), and flowering plants (C) all undergo photosynthesis, which involves the consumption of carbon dioxide and production of oxygen as a waste product.

9. **D** The early Earth lacked the large quantities of oxygen that it currently has. Methane (A), hydrogen (B), and ammonia (C) were all major components of the early Earth atmosphere.

10. **C** Protists are unicellular eukaryotic organisms which typically lack a cell wall. Bacteria (A), fungi (B), and plants (D), all use cell walls for structural stability and protection.

11. **A** In summary, homologous structures share a common ancestry; analogous structures do not.

12. **C** Disruptive selection selects for traits that are extremes rather than the norm.

13. **D** The classic example of chicks associating the first moving object that they see with their mother is called imprinting.

14. **C** Thigmotropism is the movement of plants according to physical contact with their environments. An example is the movement of ivy along surfaces. Phototropism (A) is the movement toward light. Gravitropism (B) is the movement according to gravity. Chemotropism (D) is movement due to a chemical gradient.

15. **B** Phototrophs are organisms that obtain energy by photon capture, not oxidation, so answers A and (C) are incorrect. Autotrophs are organisms that produce complex organic compounds from simple substances, so answer (D) is wrong, and (B) is correct. Answer (A) describes chemoautotrophs. Answer (C) describes chemoheterotrophs. Answer (D) describes photoheterotrophs.

16. **B** A phototropism is a movement or growth in response to light (not touch), so answer (A) is wrong. A gravitropism is a movement or growth in response to gravity, so answer (C) is incorrect. Thermotropism is a movement or growth in response to temperature, so answer (D) can be eliminated. Thigmotropism is a movement in response to touch, so answer (B) is correct.

17. **C** Habituation occurs when an animal learns not to respond to a stimulus. In answer (C), the prairie dogs learn not to respond (alarm calls) to a stimulus (exposure to humans). Thus, answer (C) is correct. Answer (A) is incorrect because it is an example of classical conditioning in which one stimulus (the bell) signals the occurrence of a second stimulus (food). Answer (B) is incorrect because it describes imprinting. Answer (D) is wrong because it describes reinforcement in which a behavior (sitting) is encouraged by a positive stimulus (treats).

18. **A** Circadian rhythm refers to an organism's internal clock. Although the internal clock affects animal behavior, the term refers to an internal mechanism, not a social behavior. Therefore, answer (A) is correct. Answers (B), (C), and (D) describe the social behavior of animals, so these answers are incorrect. Dominance hierarchies refer to the pecking orders that some communities of animals develop. Territoriality is a behavior in which animals establish and defend a certain territory. Altruistic behavior refers to unselfish behavior that benefits another organism in a group.

Chapter 29
Big Idea 2 Drill 4

BIG IDEA 2 DRILL 4

Multiple-Choice Questions

<u>Questions 1-3</u>

The sodium-potassium pump (Na^+/K^+-ATPase) is an antiporter, which pumps 3 sodium ions (Na^+) out of a cell and 2 potassium ions (K^+) into a cell. The pump is critical for maintaining the resting membrane potential and helps regulate the electrostatic environment of the cell. The sodium-potassium pump plays a critical role in re-establishing the resting membrane potential in neuron cells undergoing action potentials. A schematic of the sodium-potassium pump is shown below.

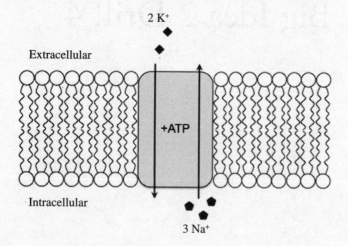

1. The sodium-potassium pump is an example of which of the following transmembrane processes?

   (A) Active transport
   (B) Simple diffusion
   (C) Facilitated diffusion
   (D) Pinocytosis

2. The sodium-potassium pump plays a critical role in re-establishing the membrane potential in a neuron following which of the following events during the action potential?

   (A) The opening of voltage-gated Na^+ channels which caused a depolarization of the neuronal membrane.
   (B) The opening of voltage-gated K^+ channels which caused a repolarization of the neuronal membrane.
   (C) The closure of voltage-gated K^+ channels following hyperpolarization of the neuronal membrane.
   (D) The closure of voltage-gated Na^+ channels following depolarization of the neuronal membrane.

3. Where are the concentrations of Na^+ and K^+ ions normally the greatest?

   (A) The Na^+ ion concentration is greatest on the inside of the cell, and the K^+ ion concentration is greatest on the outside of the cell.
   (B) The Na^+ ion concentration is greatest on the outside of the cell, and the K^+ ion concentration is inside on the outside of the cell.
   (C) The Na^+ and K^+ ion concentrations are both highest inside the cell.
   (D) The Na^+ and K^+ ion concentrations are both highest outside the cell.

4. Suppose a toxin has been applied which forms pores in the cell membrane of eukaryotic cells. Which of the following accurately describes the movement and change in charge associated with the flow of Na^+ ions?

(A) The Na^+ ions would rush out of the cell causing a depolarization of the membrane.

(B) The Na^+ ions would rush out of the cell causing a repolarization of the membrane.

(C) The Na^+ ions would rush into of the cell causing a depolarization of the membrane.

(D) The Na^+ ions would rush into of the cell causing a repolarization of the membrane.

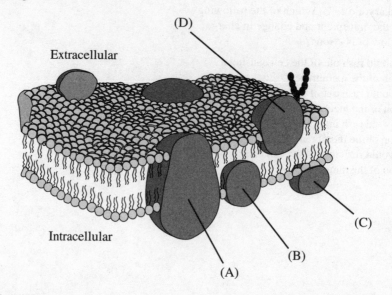

The cell membrane organization is referred to as the fluid-mosaic model due to the fluid-like flexibility and structure of the membrane. The structure above displays the plasma membrane of a cell including different types of membrane-associated proteins. Depending upon the function of the proteins, they may be present on one side of the membrane or main span the membrane. Some proteins are glycosylated, which have exposed sugar chains.

5. Which of the labeled structures in the figure above represents an integral protein?

   (A) A
   (B) B
   (C) C
   (D) D

6. Which of the labeled structures in the figure may allow for the movement of small ions such as calcium (Ca^{2+}) into the cell?

   (A) A
   (B) B
   (C) C
   (D) D

7. A glycoprotein is depicted in the model above. Which of the following accurately describes the role of these proteins?

   (A) To facilitate the transmembrane movement of small ions such as sodium or potassium down their respective concentration gradients
   (B) To facilitate adhesion and cell recognition to other cells
   (C) To increase fluidity of the plasma membrane
   (D) To serve as a receptor for cellular signaling

8. A key characteristic of the plasma membrane is its semi-permeability to substances. Which of the following may pass through the membrane by simple diffusion?

   (A) Large polar molecules
   (B) Proteins
   (C) Small ions
   (D) Small hydrophobic molecules

9. The cell pictured was placed in a hypertonic solution. Which of the following describes the movement of water, if any, across the membrane?

   (A) Into the cell
   (B) Out of the cell
   (C) Equal movement into and out of the cell
   (D) Into the space between the two layers of the phospholipid bilayer

10. The fluidity of the plasma membrane requires cytoskeletal elements to maintain shape and to serve as intracellular scaffold. All of the following are examples of cytoskeletal structures EXCEPT

    (A) microfilaments
    (B) microtubules
    (C) centrioles
    (D) endoplasmic reticulum

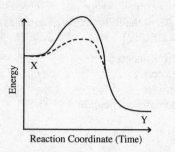

Reaction Coordinate (Time)

The image above depicts the catalyzed (dashed line) and uncatalyzed (solid line) energetics of the conversion of compound X into compound Y.

11. Which of the following best describes the net energy flow of the reaction shown above?

(A) Endergonic
(B) Exergonic
(C) Energetically neutral
(D) Endothermic

12. Which of the following best explains the role of the catalyst in the reaction shown?

(A) The catalyst lowers the energy of the reactants to increase their stability.
(B) The catalyst lowers the activation energy to decrease the energy threshold necessary to proceed from reactant to product.
(C) The catalyst lowers the energy of the product to make it more stable and efficiently produced.
(D) The catalyst absorbs the excess energy of the reaction to ensure that it remains energetically neutral.

13. Suppose the addition of ATP to the reaction increases the kinetics of the enzyme ten-fold. Which of the following would best describe the role of ATP?

(A) ATP is an allosteric inhibitor.
(B) ATP is a competitive inhibitor.
(C) ATP is a reaction substrate.
(D) ATP is a cofactor.

14. The conversion of compound X into compound Y requires a pH above 8. Which of the following describes the required conditions of the experiment?

(A) Acidic
(B) Alkaline
(C) Neutral
(D) Hydrophobic

15. In the Krebs cycle (shown below), citric acid is a 6-carbon molecule generated by the combination of acetyl-CoA with oxaloacetate. How many carbons are present in oxaloacetate?

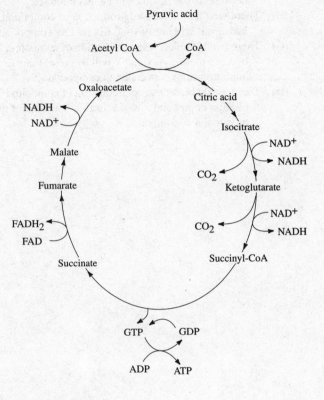

(A) 2
(B) 3
(C) 4
(D) 5

16. During the electron transport chain (ETC), protons are pumped out of the matrix of the mitochondrion and into the intermembrane space. What is the purpose of this process?

(A) To lower the pH of the inner-membrane space to activate ATP-converting enzymes present there
(B) To establish a chemiosmotic gradient to be used by an ATP synthase
(C) To remove them from the matrix, where they inhibit the activity of ATP synthase
(D) To combine them with oxygen molecules as the final electron acceptor in the intermembrane space to form water

17. "Energy cannot be created nor destroyed."

The statement above is better known as which of the following?

(A) First Law of Thermodynamics
(B) Law of Independent Assortment
(C) Second Law of Thermodynamics
(D) Law of Segregation

18. A scientist applies a chemical to a cell culture system, which removes available oxygen. Which of the following cellular processes will still occur?

    I. Glycolysis
   II. Fermentation
  III. Oxidative Phosphorylation

(A) I only
(B) I and II only
(C) II and III only
(D) I, II, and III

Questions 19-20

Negative feedback inhibition is a critical process of maintaining and control metabolic pathways. In the process, a down-stream product of the pathway cascade acts as an inhibitor of an upstream step. When the pathway fails to produce enough products, there is less inhibition. Conversely, when the pathway produces sufficient product, there is inhibition, to reduce the amount of subsequent product produced. In the following cascade, compound Y acts as an inhibitor of the conversion of compound W into compound X as shown in the schematic below.

$$W \rightarrow X \rightarrow Y \rightarrow Z$$

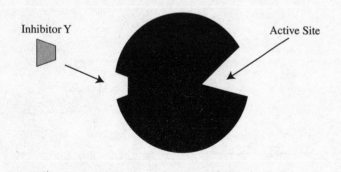

Inhibitor Y   Active Site

19. How does the binding of compound Y reduce or prevent activity of the enzyme?

(A) Compound Y is a noncompetitive inhibitor that binds at an allosteric site, which prevents binding compound W at the active site.
(B) Compound Y is a competitive inhibitor that binds at an allosteric site, which prevents binding compound W at the active site.
(C) Compound Y is a noncompetitive inhibitor that binds at the active site, which prevents binding compound W at the active site.
(D) Compound Y is a competitive inhibitor that binds at the active site, which prevents binding compound W at the active site.

20. Suppose a toxin was applied which inhibited the conversion of compound Y into compound Z. How would this effect the production of compound X?

(A) There would be increased production of compound X because its conversion will be less inhibited.
(B) There would be increased production of compound X because it will not be used to generate compound Y.
(C) There would be decreased production of compound X because more compound Y will be present due to inhibition of its conversion to compound Z.
(D) There would be decreased production of compound X because compound Z also acts as an inhibitor of the production of compound X.

Questions 21-24

Chlorophyll is a green pigment present in the chloroplasts of algae and plants. It is essential for catalyzing the light-dependent cycles in photosynthesis. A scientist purifies both forms of chlorophyll (a and b) from plant chloroplasts and evaluates them for light absorption using a spectrophotometer. The data from the spectrophotometer are shown below.

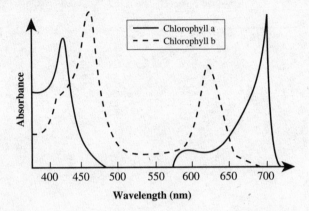

21. All of the following are products of the light-dependent cycles of photosynthesis EXCEPT

(A) ATP
(B) NADP$^+$
(C) O$_2$
(D) NADPH

22. Chlorophyll gives plants their characteristic green color. Which of the following wavelengths of light most closely represents the color green?

(A) 480 nm
(B) 540 nm
(C) 620 nm
(D) 675 nm

23. How does the pigment chlorophyll cause plants to appear green?

(A) Chlorophyll absorbs light best at wavelengths associated with green part of the visible spectrum.
(B) Chlorophyll absorbs red and blue/violet lights and reflects mostly green wavelengths of light.
(C) Chlorophyll absorbs violet light and uses some of the energy in the generation of ATP and NADPH, the remaining energy is released as a lower energy green photon.
(D) Chlorophyll absorbs red light and uses some of the energy in the generation of ATP and NADPH, the remaining energy is released as a lower energy green photon.

24. Most plants undergo non-cyclic photophosphorylation rather than cyclic. Why?

(A) Cyclic photophosphorylation fails to produce ATP, instead producing lower energy ADP molecules.
(B) Cyclic photophosphorylation fails to produce NADPH, which is necessary for the light-independent pathway.
(C) Cyclic photophosphorylation fails to recycle CO$_2$ gas generated as a byproduct of cellular respiration in the mitochondria.
(D) Cyclic photophosphorylation requires more energy input than noncyclic photophosphorylation.

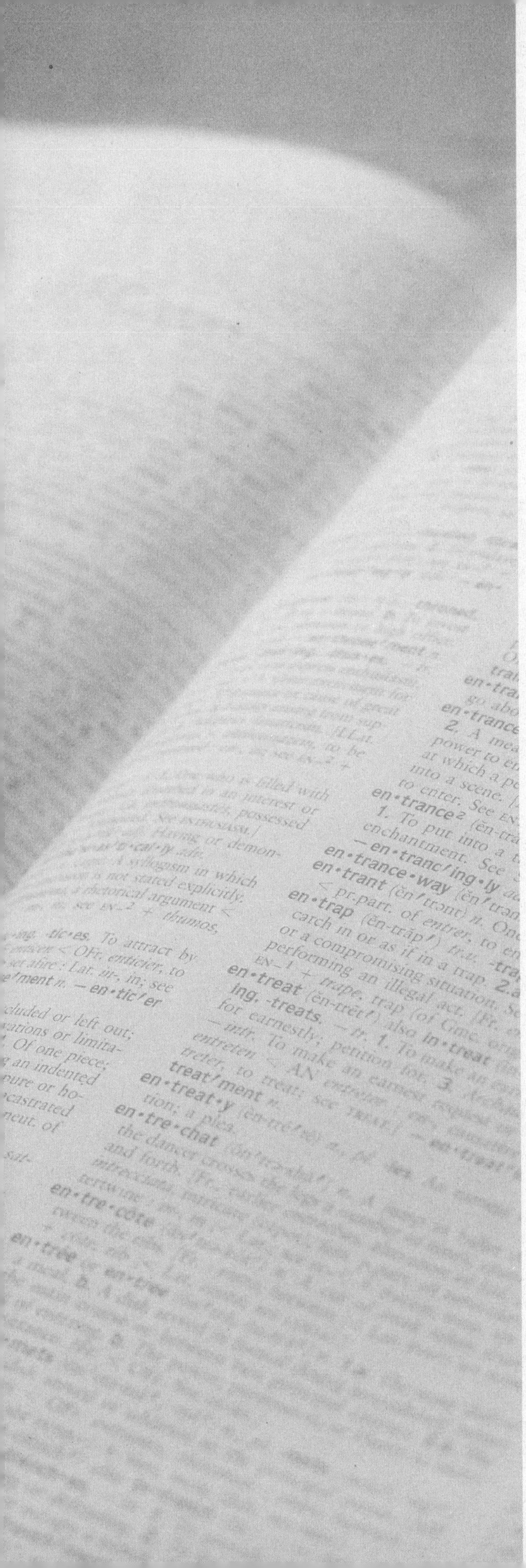

Chapter 30
Big Idea 2
Drill 4 Answers and Explanations

ANSWER KEY

1. A
2. C
3. B
4. C
5. A
6. A
7. B
8. D
9. B
10. D
11. B
12. B
13. D
14. B
15. C
16. B
17. A
18. B
19. A
20. C
21. B
22. B
23. B
24. B

ANSWERS AND EXPLANATIONS

Multiple-Choice Questions

1. **A**   The sodium-potassium pump is an example of active transport. In this process, materials are moved across the cell membrane against their respective gradients through the use of energy (ATP). Diffusion involves the movement of substances across the cell membrane from concentrations of high to low, either with the need of an assisting protein (C) or without (B). Pinocytosis (D) is the cellular process by which a cell intakes fluids.

2. **C**   In the neuronal cell, the sodium-potassium pump plays a key role in re-establishing the resting membrane potential following the closure of the voltage-gated K^+ channels.

3. **B**   The concentrations of the sodium and potassium ions are greatest outside and inside of the cell, respectively. One way to figure this out is to consider which way the sodium-potassium pump is having to "pump" the ions. For example, the potassium is being pushed against its gradient (it wants to leave the cell; high to low).

4. **C**   The opening of a pore would have the same effect as an action potential in a neuron. The sodium ions, which are in high concentration outside of the cell, would rush in, causing a depolarization of the membrane.

5. **A**   Integral proteins are proteins, which span the membranes, are exposed to both the intracellular and extracellular surfaces. Choices (B) and (C) represent peripheral proteins. (D) is an example of a glycoprotein.

6. **A**   Ions require a channel or pump to facilitate their movement across membranes. Only the integral protein (A) spans both membranes.

7. **B**   Glycoproteins are exposed to the extracellular environment and facilitate adhesion (by glycan interactions) and cell recognition.

8. **D**   Simple diffusion is the movement of small molecules across the membrane from high concentrations to low concentrations without the need of a channel or pore. Small hydrophobic molecules (D) are capable of crossing the membrane without assistance. Polar molecules such as proteins (B) and ions (C) require the assistance of ion channels or pores to cross the phospholipid bilayer.

9. **B**   Hypertonic solutions are those that have a higher concentration of salt relative to the inside of the cell. Because water undergoes osmosis from high concentrations to low concentrations, water will move from the inside of the cell to the outside (where there is less water and more salt).

10. **D** The endoplasmic reticulum is an intracellular structure involved in the synthesis of proteins and plays a part in the secretory pathway. Microfilaments (A), microtubules (B), and centrioles (C) are all cytoskeletal structures.

11. **B** The reaction shown is an example of an exergonic process because the reactants have higher energy than the products. The surplus energy is released in an exergonic reaction.

12. **B** Catalysts speed up reactions by lowering the activation energy, which is the energy threshold required to proceed from reactants to products. Catalysts do not alter the energy of the reactants (A) or products (C) and do not absorb excess energy (D).

13. **D** ATP serves as a cofactor, a chemical that increases the activity of an enzyme. Because it increases the enzyme activity it does not act as an inhibitor and is not a substrate for the reaction.

14. **B** The enzyme requires a pH above 7, which is considered basic or alkaline.

15. **C** Because acetyl-CoA and oxaloacetate react to form citric acid (a six-carbon molecule). Between citric acid and oxaloacetate, two carbon dioxide molecules are produced, which means that oxaloacetate must have two less carbons or a total of four carbons.

16. **B** The movement of hydrogen ions into the intermembrane space of the mitochondrion is to generate a chemiosmotic gradient which can be used to generate ATP using ATP synthase.

17. **A** The first law of thermodynamics states that energy can neither be created nor destroyed.

18. **B** Glycolysis and fermentation do not require oxygen to occur and typically constitute the sole ATP generated pathways in anaerobic conditions for many cells. Oxidative phosphorylation, as its name implies, requires oxygen as a final electron acceptor to generate ATP in the electron transport chain (ETC).

19. **A** Compound Y acts as a noncompetitive inhibitor because it does not compete with compound W for the active site of the enzyme, rather it binds to a different site (allosteric site) and inactivates the enzyme.

20. **C** Because compound Y inhibits the generation of compound X, blocking the conversion of compound Y into compound Z will result in a relative increase in compound Y produced and therefore greater inhibition of the generation of compound X.

21. **B** ATP, O_2, and NADPH are all produced during the light-dependent cycles. ATP and NADPH are used in the light-independent cycles to generate sugar from carbon dioxide. $NADP^+$ is produced as a byproduct of the light-independent cycles.

22. **B** Because plants have chlorophyll, their color is due to the reflection of light (the lack of absorption) in the green part of the color spectrum. For both types of chlorophyll, the minima fall around 540 nm.

23. **B** Chlorophyll (as shown in the figure) absorbs red and blue wavelengths of light and reflects the color green.

24. **B** Cyclic photophosphorylation fails to produce NADPH, which is a necessary substrate for the light-independent cycles.

Chapter 31
Big Idea 3 Drill 4

BIG IDEA 3 DRILL 4

Multiple-Choice Questions

Questions 1-4

Hemophilia is a genetic, X-linked disease, which causes disruption in the blood-clotting cascade. People affected by the disease experience bruising and in severe cases may bleed to death due to the inability to form clots. The disease almost exclusively affects males, however in very rare situations, women may also have the disease. A hemophilic man marries a woman whose father was also a hemophilic. The family consults a geneticist about the chances of their children having the disease.

1. What is the chance that their first son will have the disease?

   (A) 25%
   (B) 50%
   (C) 75%
   (D) 100%

2. Is it possible for the couple to have a daughter affected with the disease?

   (A) Yes, because both parents carry the diseased X chromosome.
   (B) Yes, because the child's father carries the diseased X chromosome.
   (C) No, because only the child's father is affected by the disease.
   (D) No, because women cannot get hemophilia.

3. Why is it extremely rare for women to acquire the disease?

   (A) Women would not survive menstruations during puberty and thus are unable to have children to pass on the disease.
   (B) Women would require two copies of the defective gene because they have two copies of their X chromosome.
   (C) Women utilize different clotting factors than men due to their physiologic needs during childbirth.
   (D) Women are less susceptible to the pathogen, which causes the disease.

4. What is the probability that couple's first child will be affected by the disease? Provide your answer as a fraction.

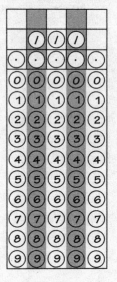

5. A woman that has a B blood type is involved in a car accident and requires an emergency blood transfusion. Which of the following donor bloods can she receive?

    I. A type
   II. B type
  III. AB type
  IV. O type

(A) II only
(B) I and III only
(C) II and III only
(D) II and IV only

6. A test cross is performed between pea plants for two traits. However, when the progeny generation is evaluated, the ratio of traits differs substantially from all expected ratios based on Mendelian genetics. Which of the following may best explain this phenomenon?

(A) The traits are codominant, and therefore Mendelian genetics do not apply.
(B) The traits are linked, and therefore independent assortment did not occur.
(C) The traits represent cases of incomplete dominance.
(D) The traits are epistatic upon a third, unidentified trait.

7. A new toxin has been identified which causes single nucleotide mutations during DNA replication. Which of the following changes would be the most deleterious effect on the production of a protein?

(A) The nucleotide substitution resulted in a silent mutation at the beginning of the exon region of a gene.
(B) The nucleotide substitution resulted in a nonsense mutation near the beginning of the exon region of a gene.
(C) The nucleotide substitution resulted in a missense mutation near the beginning of the exon region of a gene.
(D) The nucleotide substitution resulted in a missense mutation near the end of the exon region of a gene.

8. During DNA replication, which of the following enzymes synthesizes RNA?

(A) Ligase
(B) Primase
(C) DNA polymerase
(D) Helicase

9. If a messenger RNA has the sequence AAG, which of the following would be the complementary anticodon triplet in a tRNA?

(A) TTC
(B) CTT
(C) UUC
(D) AUG

10 Which of the following represents the maximum number of amino acids in a protein whose gene is encoded by 60 nucleotides?

(A) 20
(B) 60
(C) 120
(D) 180

Questions 11-13

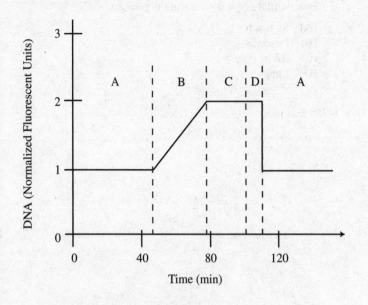

An experiment is performed to evaluate the amount of DNA present during a complete cell cycle. All of the cells were synced prior to the start of the experiment. During the experiment, a fluorescent chemical was applied to cells, which would only fluoresce when bound to DNA. The results of the experiment are shown above. Differences in cell appearance by microscopy or changes in detected DNA were determined to be phases of the cell cycle and are labeled with the letters A through D.

11. Which of the phases above represents the S phase of the cell cycle?

(A) Phase A
(B) Phase B
(C) Phase C
(D) Phase D

12. What is happening during Phase D?

(A) The genomic DNA is being replicated.
(B) The cell is generating more resources in preparation for cell division.
(C) The cell is undergoing cell division.
(D) The cell is interphase and undergoing daily activities.

13. How long did the G2 phase of the cell cycle last for the cell evaluated above?

(A) 15 min
(B) 25 min
(C) 40 min
(D) 60 min

14. Which of the following processes occurs in the nucleus of a eukaryotic cell?

    (A) Translation
    (B) Transcription
    (C) Conjugation
    (D) Exocytosis

16. Which of the following tissues is not derived from the endodermal germ layer during organogenesis?

    (A) Stomach
    (B) Pancreas
    (C) Brain
    (D) Lungs

15 Which of the following structures is where spermatogonia undergo meiosis?

    (A) Ovaries
    (B) Seminiferous tubules
    (C) Interstitial cells of the testes
    (D) Prostate gland

17. Which of the following is the proper order of embryological development?

(A) Zygote, Morula, Blastula, Gastrula, Organogenesis
(B) Zygote, Blastula, Morula, Gastrula, Organogenesis
(C) Zygote, Blastula, Morula, Organogenesis, Gastrula
(D) Zygote, Morula, Blastula, Organogenesis, Gastrula

18. Two flowering plants are crossed, where short height (S) is dominant over tall height (s), and red flower color (R) is dominant over white flower color (r). In the cross, a flower that is a heterozygote for both traits is crossed with a flower that is recessive for both traits. Approximately what percentage of the progeny generation is expected to be short red flowers?

(A) 25%
(B) 50%
(C) 75%
(D) 100%

Chapter 32
Big Idea 3
Drill 4 Answers and Explanations

ANSWER KEY

1. B
2. A
3. B
4. $\dfrac{1}{2}$
5. D
6. B
7. B
8. B
9. C
10. A
11. B
12. C
13. B
14. B
15. B
16. C
17. A
18. A

ANSWERS AND EXPLANATIONS

Multiple-Choice Questions

1. **B** Because the father has to pass his Y chromosome onto his son, the transmission of hemophilia is completely dependent on the mother. Because the mother's father had hemophilia, she had to acquire a bad copy of the X chromosome. Therefore, she is a carrier and has a 50% chance of passing on the diseased allele to her son.

2. **A** Because the father has a diseased X chromosome and the mother is a carrier, there is a chance (50%) that they will have a daughter with two diseased X chromosomes.

3. **B** The rarity of X-linked diseases in women is due to the fact that they have two copies of the X chromosome and thus typically would require two diseased alleles.

4. $\frac{1}{2}$ The probability that their first child will be affected is the same whether it is male or female (fifty-fifty chance he acquires the diseased X from his mother).

5. **D** Individuals with a B blood type have B antigen on their cells. They would therefore have antibodies against the A antigen. Because blood types A and AB both exhibit A antigen, they are incompatible with B blood. O blood lacks surface antigens and therefore would not result in an immune reaction.

6. **B** Deviations in the expected genetic ratios are often due to linkage (or non-independent assortment of genes). If the genes were codominant (A), both traits would co-expressed. In cases of incomplete dominance (C), there would be blending of traits.

7. **B** Nonsense mutations cause the introduction of a stop codon. Because this mutation will be located near the beginning of the exon (coding region of the gene), the protein will be significantly shortened from its original form. Missense mutations result in the coding of a different amino acid, however, the rest of the protein sequence will be translated normally.

8. **B** During DNA replication, the enzyme primase synthesizes short RNA primers, which are used by the DNA polymerase as templates to begin complementary DNA synthesis in the 5' to 3' direction. Ligase (A) anneals DNA segments near the end of the DNA replication, and the helicase (D) is responsible for unwinding the DNA prior to DNA polymerization.

9. **C** In RNA pairing, adenines (A's) pair with uridines (U's) and cytosines (C's) pair with guanines (G's). Because answers (A) and (B) lack uridines instead of thymines, they cannot be the correct sequence. The correct pairing would be UUC.

10.   **A**   Each codon (set of 3 nucleotides) codes for 1 amino acid. Assuming the entire gene is translated, the maximum number of amino acids translated for a gene consisting of 60 nucleotides would be 20.

11.   **B**   During the S phase (or synthesis phase) of the cell cycle, the DNA concentration should increase as the cell prepares a complete copy of its genetic material for cell division. Only phase B displays a net increase in the amount of detected DNA.

12.   **C**   During phase D, the amount of detected DNA per cell is decreasing, which is indicative of cell division (the same amount of DNA spread over more cells results in a net reduction of DNA per cell). Phases A and C represent G1 and G2, where the cell is in interphase and is undergoing daily activities (D), including generating more resources for cell division (B). Phase B represents the S phase, where the genomic DNA is copied (A).

13.   **B**   The G2 phase appears to last from approximately the 75 min mark to the 100 min mark, or a net time of 25 min.

14.   **B**   Translation is the process of converting mRNA to a polypeptide. Translation occurs in the cytoplasm or in an organelle such as the rough endoplasmic reticulum. Translation does not occur in the nucleus of eukaryotic cells, so answer (A) is wrong. Conjugation is a transfer of genetic material between bacterial cells, but bacterial cells are not eukaryotic, so answer (C) is wrong. Exocytosis is a process in which cells secrete substances, and these processes do not occur in the nucleus, so answer (D) is wrong. Transcription is the process of synthesizing mRNA from a DNA template. Because the genetic material of a eukaryotic cell is in the nucleus, transcription must take place in the nucleus. Thus, answer (B) is correct.

15.   **B**   The seminiferous tubules of the testes are the site of meiosis for spermatogonia. The interstitial cells (C) are responsible for producing and secreting testosterone. The prostate gland (D) produces fluids, which become part of semen prior to ejaculation. Ovaries (A) are not part of male anatomy.

16.   **C**   The endoderm gives rise to GI tract and respiratory tract tissues and organs. The brain and nervous system are derived from the ectoderm.

17.   **A**   The correct order of development is zygote (fertilized egg), morula (ball of divided cells), blastula (hollow sphere of divided cells), gastrula (gastrulation occurs), and organogenesis (formation of germ layers and start of organ tissues).

18.   **A**   A heterozygote for both traits (SsRr) is crossed with a homozygote for both traits (ssrr). There is a 50% chance that red flowers are observed and a 50% chance that short flowers are observed. Therefore, there is a 25% chance that the progeny generation will be short, red flowers.

Chapter 33
Big Idea 4 Drill 4

BIG IDEA 4 DRILL 4

Multiple-Choice Questions

Questions 1-5 refer to the following paragraph

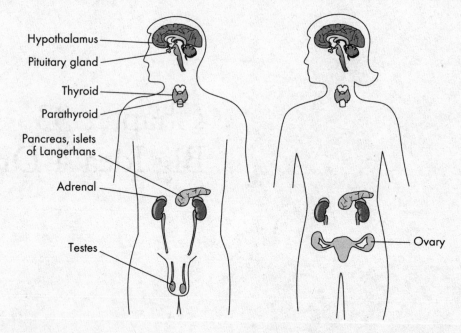

Hypothalamus
Pituitary gland
Thyroid
Parathyroid
Pancreas, islets of Langerhans
Adrenal
Testes
Ovary

The endocrine system is comprised of a series of glands (as shown above), which produce hormones that maintain homerostasis in the body. Two key hormones in the body are calcitonin and parathyroid hormone (PTH). These hormones regulate the calcium ion concentrations in the blood by controlling bone remodeling. Calcium is vital to various cellular responses, and under conditions of low calcium concentrations in the blood, will trigger bone dissolution to release more calcium in the blood.

1. During conditions of low calcium concentrations in the blood, which of the hormones is most likely to be released?

   (A) Calcitonin
   (B) Parathyroid hormone
   (C) Growth hormone
   (D) TSH

2. Which of the glands shown above is responsible for synthesizing calcitonin?

   (A) Adrenal glands
   (B) Thyroid gland
   (C) Pituitary gland
   (D) Hypothalamus

3. Calcitonin has been released due to abrupt changes in serum calcium concentrations. Its activity will most likely trigger bone activity by which of the following cells?

    (A) Osteocytes
    (B) Osteoblasts
    (C) Osteoclasts
    (D) Chrondrocytes

4. Which type of bone cells is responsible for building bone from calcium and phosphate salts?

    (A) Osteocytes
    (B) Osteoblasts
    (C) Osteoclasts
    (D) Chrondrocytes

5. The pituitary gland is considered a master gland because it secretes many hormones, including hormones, which regulate the release of other hormones. All of the following are hormones secreted by the pituitary gland EXCEPT

    (A) prolactin
    (B) thyroid-stimulating hormone (TSH)
    (C) follicle-stimulating hormone (FSH)
    (D) epinephrine

Diabetes is a disease associated with high concentrations of glucose in the blood. The body uses several hormones to regulate serum sugar concentrations including glucagon and insulin. In type I diabetes, individuals are unable to produce insulin. In type II diabetes, individuals suffer from insulin resistance. In both cases, consumption of foods high in sugars are carefully watched to limit dangerously high sugar concentrations.

6. Where are the hormones glucagon and insulin produced in the body?

    (A) Adrenal glands
    (B) Pituitary gland
    (C) Thyroid gland
    (D) Pancreas

7. Which form of diabetes is more common in young children?

    (A) Type I, because they are born with insulin deficiency
    (B) Type I, because they have acquired insulin deficiency due to high sugar diets
    (C) Type II, because high intake of sugars has cause tolerance of the insulin hormone and cells no longer respond to its presence
    (D) Type II, because they lack the same amount of receptors for the hormone insulin

8. Which type of cells in the pancreas is responsible for synthesizing insulin?

    (A) Alpha cells
    (B) Beta cells
    (C) Gamma cells
    (D) Delta cells

Questions 9-14 refer to the following paragraph.

The menstrual cycle plays a critical role in regulating the formation and release of a follicle from the ovaries and preparation and maintenance of the uterus for its fertilization and implantation. The menstrual cycle lasts approximately 28 days and consists of three main phases: the follicular phase (when follicle matures and the ovum is released in a process called ovulation), the luteal phase (when the ruptured follicle remaining in the ovary becomes the corpus luteum), and the menstrual phase (when the endometrial lining of the uterus is sloughed off to reset the uterine lining to restart the cycle).

9. The rupture of the mature follicle during ovulation is directly caused by a surge of which of the following hormones?

   (A) Luteinizing hormone (LH)
   (B) Estrogen
   (C) Follicle-stimulating hormone (FSH)
   (D) Progesterone

10. Following ovulation, the corpus luteum begins producing which of the following hormones?

   (A) Luteinizing hormone (LH)
   (B) Estrogen
   (C) Follicle-stimulating hormone (FSH)
   (D) Progesterone

11. Which of the following hormones plays an important role in helping the endometrium grow following menstruation?

   (A) Luteinizing hormone (LH)
   (B) Estrogen
   (C) Follicle-stimulating hormone (FSH)
   (D) Progesterone

12. Which of the following hormones is responsible for the surge in LH prior to ovulation?

   (A) Testosterone
   (B) Estrogen
   (C) Follicle-stimulating hormone (FSH)
   (D) Progesterone

13. How does the body know to transition from the luteal phase to the menstrual phase?

   (A) The implanted embryo produces human chorionic gonadotropin (hCG) which triggers menstruation.
   (B) The endometrium of the uterus is under strict control by melatonin. When melatonin levels fall, menstruation is triggered.
   (C) The endometrium is maintained by progesterone released by the corpus luteum. As the corpus luteum degrades, decreases in progesterone trigger menstruation.
   (D) The pituitary gland secretes luteinizing hormone which triggers menstruation after receiving a signal from the uterus of no implantation of the embryo.

14. Approximately how long does the luteal phase last before the start of menstruation?

   (A) 5 days
   (B) 1 week
   (C) 2 weeks
   (D) 4 weeks

15. Leukocytes play a role in which of the following functions of human physiology?

   (A) Transportation of oxygen
   (B) Blood clotting
   (C) Immunity and protection from foreign organisms
   (D) The conversion of CO_2 to bicarbonate

16. Which of the following hormones is primarily responsible for phototropism in plants?

   (A) Auxin
   (B) Cytokinin
   (C) Gibberellin
   (D) Ethylene

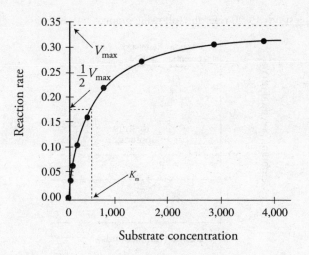

17. The diagram above is a saturation curve for a chemical reaction that is catalyzed by an enzyme. What substrate concentration corresponds to a reaction rate less than $\frac{1}{2}V_{max}$?

   (A) 200
   (B) 800
   (C) 1500
   (D) 3300

18. Which of the following digestive enzymes catalyzes the breakdown of protein?

   (A) Salivary amylase
   (B) Pancreatic amylase
   (C) Pancreatic lipase
   (D) Chymotrypsin

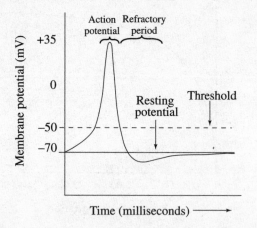

19. At +35 mV, a sudden change occurs in the membrane potential. Which of the following best describes why this change occurs.

    (A) The voltage-gated sodium channels open, allowing sodium to enter the cell causing depolarization.
    (B) The voltage-gated potassium channels open, allowing potassium to enter the cell causing depolarization.
    (C) The voltage-gated sodium channels open, allowing potassium to leave the cell causing repolarization.
    (D) The voltage-gated potassium channels open, allowing potassium to leave the cell causing repolarization.

During a neural transmission, neurons release neurotransmitters that bind to ligand-gated receptors on neighboring neurons inducing an action potential. During an action potential, a wave of depolarization travels down the length of the axon where it reaches the axon terminals. Depolarization of the axon terminals results in the release of neurotransmitters and the cascade continues with downstream neurons. Shown above is a diagram depicting an action potential.

20. Why is there a refractory period during which no action potentials may occur following the wave of depolarization?

    (A) The voltage-gated potassium channels are unable to open and thus potassium may not leave the cell during the refractory period.
    (B) The voltage-gated sodium channels are unable to open and thus sodium may not enter the cell during the refractory period.
    (C) The membrane potential is too low to allow for the threshold to be met during the refractory period.
    (D) The voltage-gated potassium channels are unable to close and thus potassium continues to leave the cell preventing depolarization.

21. A hiker has been recently bitten by a snake and has experienced a robust parasympathetic response due to the snake venom. Which of the following physiological changes is likely to occur?

    (A) Constriction of blood vessels
    (B) Elevated heart rate
    (C) Increase in respiration
    (D) Increased digestion

22. A motorcyclist was involved in an accident and has sustained brain damage causing loss of muscle coordination and refined motion. Which of the following regions of the brain has likely been damaged?

    (A) Cerebrum
    (B) Cerebellum
    (C) Hypothalamus
    (D) Midbrain

23. An environment is associated with short coniferous trees, long winters, and animal life consisting of caribou, moose, bears, and rabbits. Which of the following biomes best describes this environment?

    (A) Grasslands
    (B) Deciduous Forest
    (C) Tundra
    (D) Taiga

24. In a pond ecosystem, spring rains trigger an expansion of species at levels of the food chain. Runoff from nearby hills brings nutrients which, when combined with warming temperatures, trigger an algae bloom. The populations of small protozoans such as plankton expand by ingesting the algae. Subsequently, these plankton are consumed by small crustaceans such as crayfish, which ultimately become prey for fish such as catfish or carp. In this ecosystem, which of the following accurately describes the role of the plankton, as described?

    (A) They are producers.
    (B) They are primary consumers.
    (C) They are secondary consumers.
    (D) They are tertiary consumers.

25. An individual has recently survived a myocardial infarction (heart attack) resulting in partial cell death in the muscle surrounding the left ventricle. This individual's heart would have difficulty pumping blood directly to which of the following?

    (A) Left atrium
    (B) Right atrium
    (C) Pulmonary artery and the lungs
    (D) Aorta and rest of the body

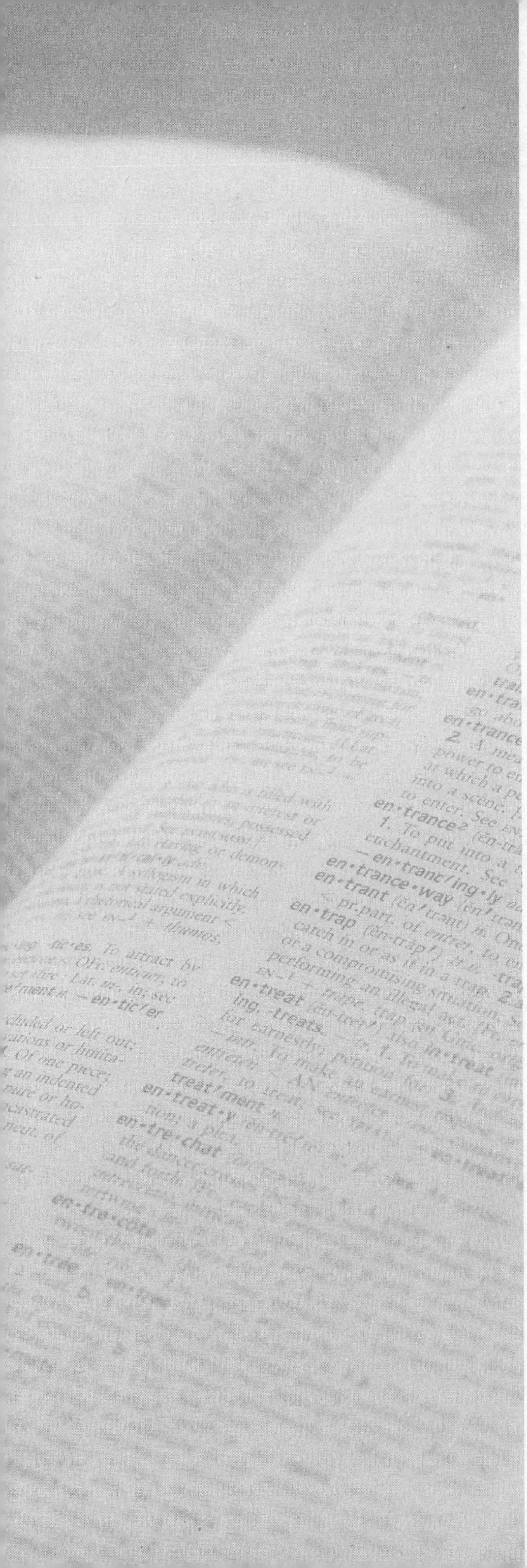

Chapter 34
Big Idea 4
Drill 4 Answers and Explanations

ANSWER KEY

1. B
2. B
3. B
4. B
5. D
6. D
7. A
8. B
9. A
10. D
11. B
12. B
13. C
14. C
15. C
16. A
17. A
18. D
19. C
20. B
21. D
22. B
23. D
24. B
25. D

ANSWERS AND EXPLANATIONS

Multiple-Choice Questions

1. **B** Parathyroid hormone is released during conditions of low serum calcium to increase the blood calcium levels.

2. **B** Calcitonin is synthesized and secreted by the thyroid gland.

3. **B** Calcitonin decreases serum calcium levels. One key way of doing so is to increase its use in bone formation by osteoblasts.

4. **B** Osteoblasts are responsible for building bone, whereas osteoclasts (C) are responsible for breaking down bone. Osteocytes (A) are bone cells and chondrocytes (D) form cartilage.

5. **D** Epinephrine is synthesized and secreted by the adrenal glands. Prolactin (A), TSH (B), and FSH (C) are all secreted by the pituitary gland.

6. **D** Glucagon and insulin are produced by the alpha and beta cells, respectively, of the pancreas.

7. **A** Type I diabetes is sometimes referred to as juvenile diabetes because it is typically present at or near birth. Type II diabetes is acquired due to tolerance of insulin in the diet.

8. **B** The beta cells of the pancreas are responsible for synthesizing and secreting insulin.

9. **A** A rapid surge in the release of luteinizing hormone (LH) triggers the mature follicle to burst and release an ovum.

10. **D** Following corpus luteum formation, progesterone is produced (in addition to estrogen) and maintains the endometrium for implantation.

11. **B** Estrogen plays a key role in helping the endometrium grow following menstruation.

12. **B** As the follicle matures, it process increasingly greater amounts of estrogen. When the maturation is near complete a high level of estrogen triggers the LH surge that causes ovulation.

13. **C** The progesterone produced by the corpus luteum protects the endometrium from degradation. When the corpus luteum dies and degrades, the amount of progesterone declines rapidly, and the uterus enters menstruation.

14. **C** The endometrial lining is maintained approximately 2 weeks by progesterone secretions from the corpus luteum. Because ovulation occurs mid cycle and a cycle is approximately 4 weeks, the luteal phase is nearly two weeks.

15. **C** Leukocytes (white blood cells) play a role in immunity and protection from foreign organisms. Therefore, answer (C) is correct. Answer (A) describes the role of erythrocytes (red blood cells). Answer (B) describes the role of platelets. Answer (D) describes the function of the enzyme carbonic anhydrase.

16. **A** Phototropism is a growth or movement in response to light. Auxin is a major plant hormone that serves many functions. Auxin promotes growth on one side of a plant, making the plant bend toward the light, so answer (A) is correct. Cytokinin promotes cell division and differentiation, but it is not responsible for phototropism, so answer (B) is incorrect. Gibberellin promotes stem elongation, but it is not involved in phototropism, so answer (C) is incorrect. Ethylene induces leaf abscission, but it is not involved in phototropism, so answer (D) can be eliminated.

17. **A** The reaction rate $\frac{1}{2} V_{max}$ is about 17.5, and this corresponds to a substrate concentration around 600. As the saturation curve bends down from $\frac{1}{2} V_{max}$, it also bends to the left. That is, lowering the reaction rate entails lowering the substrate concentration. So the substrate concentration for a reaction rate less than $\frac{1}{2} V_{max}$ would have to be less than 600. Answer (A) is the only answer choice with a figure less than 600, so (A) is correct.

18. **D** An amylase (whether salivary or pancreatic) breaks down starch, so answers (A) and (B) can be eliminated. A lipase catalyzes the break down of fat (lipids), so answer (C) is incorrect. That leaves answer (D); trypsin and chymotrypsin are pancreatic enzymes that break down protein.

19. **C** At +35 mV, the voltage-gated sodium channels close and voltage-gated potassium channels open, which allow potassium ions to rush out of the cell. In leaving the cell, the cell membrane potential becomes more negative (repolarizes).

20. **B** A new action potential requires the reopening of the voltage-gated sodium channels. Following a wave of depolarization, these channels cannot open until the voltage-gated potassium channels close. Therefore, there is an absolute refractory period following the closure of the voltage-gated sodium channels and until the voltage-gated potassium channels close.

21. **D** Parasympathetic responses are associated with "rest and digest." Constriction of blood vessels (A), elevated heart rate (B), and increased respiration (C) are all indicative of sympathetic responses ("fight or flight").

22. **B** The cerebellum is responsible for coordinated motion and balance. Damage to this region would cause a loss of these fine motor skills.

23. **D** Taigas are located in northerly latitudes and consist of short conifers and long winters. The presence of trees denotes a key difference from tundras.

24. **B**   Primary consumers consume the producers in the biosphere. Because plankton consume algae, they are considered primary consumers in this biosphere.

25. **D**   Muscle death in the left ventricle would lead to difficulty pumping blood into the aorta and thus the rest of the body. Keep in mind: The right ventricle is responsible for pumping blood to the lungs via the pulmonary artery for oxygenation.

Chapter 35
Free-Response Short
Answers Drill

FREE-RESPONSE SHORT ANSWERS DRILL

1. Describe the driving force of evolution.

   a. List three different areas of science that provide evidence for evolution.

      _____

      _____

      _____

      _____

      _____

      _____

      _____

      _____

      _____

      _____

   b. Give an example from each of these three areas providing evidence for evolution.

      _____

      _____

      _____

      _____

      _____

      _____

      _____

      _____

      _____

      _____

2. Explain the difference between and provide examples of how natural selection occurs internally and externally.

_____

_____

_____

_____

_____

_____

_____

_____

_____

_____

3. Explain what it means for a gene to be sex-linked.

    a. Describe how you would identify that an abnormal trait was due to a sex-linked gene.

_____

_____

_____

_____

_____

_____

_____

_____

_____

_____

b. Discuss the difference between sexes in expressing an abnormal sex-linked gene.

_____

_____

_____

_____

_____

_____

_____

_____

_____

_____

4. Describe the process of fermentation.

a. Give examples of some organisms that rely on fermentation for energy.

_____

_____

_____

_____

_____

_____

_____

_____

_____

_____

b. Explain when and why human muscle cells undergo fermentation.

_____

_____

_____

_____

_____

_____

_____

_____

_____

_____

5. Explain if viruses are considered living or nonliving and describe why. Describe the two main different life cycles of viruses.

_____

_____

_____

_____

_____

_____

_____

_____

_____

_____

_____

6. Describe the flow of energy through an ecosystem and discuss how energy is transferred from one level to the next.

_____

_____

_____

_____

_____

_____

_____

_____

_____

_____

7. Explain the technique of polymerase chain reaction (PCR), and give an example of why you might employ this technique.

_____

_____

_____

_____

_____

_____

_____

_____

_____

_____

8. Describe the different phases of the bacterial growth curve below and explain the underlying processes for each phase.

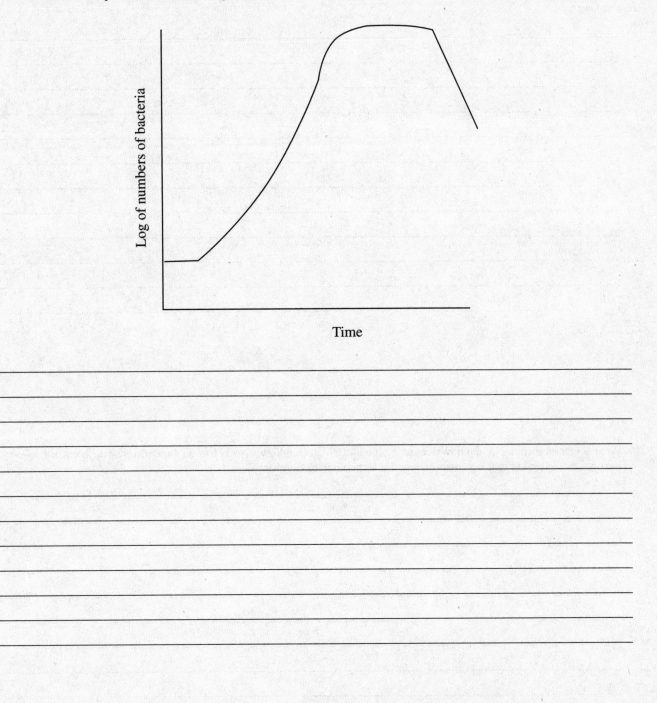

Time

_____

_____

_____

_____

_____

_____

_____

_____

_____

_____

_____

_____

9. Give an example of a system that utilizes hormonal negative feedback in order to transmit information which will then be used in regulation.

_____

_____

_____

_____

_____

_____

_____

_____

_____

_____

10. What is recombinant DNA and how does recombinant DNA technology work? How do similarities or differences between eukaryotic and prokaryotic organisms allow this technology to be utilized?

_____

_____

_____

_____

_____

_____

_____

_____

_____

_____

11. Explain the similarities and differences between the three types of muscle tissues, and give an example of a specialization that helps with function.

_____

_____

_____

_____

_____

_____

_____

_____

_____

_____

_____

12. In an experiment, a biologist grew two bacterial colonies at two different temperatures, and measured their oxygen consumption over two hours. Results are shown in the graph below.

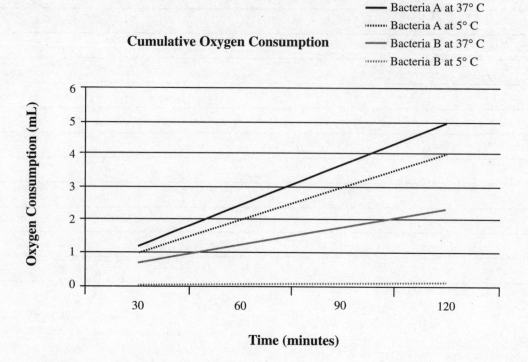

Cumulative Oxygen Consumption

— Bacteria A at 37° C
·········· Bacteria A at 5° C
— Bacteria B at 37° C
·········· Bacteria B at 5° C

Calculate the rate of oxygen consumption for each type of bacteria at each temperature. What does the rate of oxygen consumption imply about each bacteria's metabolic activity? Discuss how this is affected by temperature for each bacteria.

_____

_____

_____

_____

_____

_____

_____

_____

_____

_____

_____

_____

13. Design an experiment to measure the effects of pH and temperature on the activity of peroxidase on the following reaction: $2 H_2O_2 \rightarrow 2 H_2O + O_2$. Be sure to include a hypothesis and how you will record data to determine if your hypothesis is true or false.

_____

_____

_____

_____

_____

_____

_____

_____

_____

_____

_____

14. Populations of a bird species are found living in mountainous regions at almost 3,000 meters above sea level. Other populations of birds that seem to be very similar to those birds living in the mountains have been seen in the same region but in valleys that are 2,000 meters below sea level. Describe a way you could use to determine if the two different populations are actually the same species of bird or if they represent distinct species.

_____

_____

_____

_____

_____

_____

_____

_____

_____

_____

15. Describe three important differences between prokaryotic and eukaryotic cell parts, and explain how these differences affect cellular processes in each type of cell.

_____

_____

_____

_____

_____

_____

_____

_____

_____

_____

_____

16. Living organisms depend on water to maintain cellular structure and function.

    a. Name the two fundamental properties of water molecules that contribute to its other special properties.

_____

_____

_____

_____

_____

_____

_____

_____

_____

_____

_____

    b. Identify and explain three of these other special properties.

_____

_____

_____

_____

_____

_____

_____

_____

_____

_____

_____

_____

17. Explain how different genes are expressed during different periods over the lifespan of an organism and how this gene expression is regulated.

_____

_____

_____

_____

_____

_____

_____

_____

_____

_____

_____

18. A population of microscopic eukaryotic organisms is grown in a flask, with the resulting growth curve as shown below.

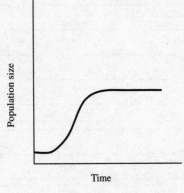

a. Explain the factors that cause the initial positive slope and then the flat portion of the curve.

_____

_____

_____

_____

_____

_____

_____

_____

_____

_____

_____

_____

b. Predict what would happen if phosphate is added to the flask somewhere along the flat portion of the curve.

_____

_____

_____

_____

_____

_____

_____

_____

_____

_____

_____

19. a.   Describe the cell cycle, including the different characteristics of each phase.

_____

_____

_____

_____

_____

_____

_____

_____

_____

_____

_____

b.   Explain how the cell cycle is regulated.

_____

_____

_____

_____

_____

_____

_____

_____

_____

_____

_____

20. Compare and contrast the mechanisms of action between steroid hormones and protein hormones, and give one example of each.

_____

_____

_____

_____

_____

_____

_____

_____

_____

_____

_____

21. While single base-pair mutations may sometimes not have an effect on structure/function of a protein, some genetic problems, such as sickle-cell anemia and cystic fibrosis, result from such mutations.

a. Explain how these mutations may alter structure and function of proteins.

_____

_____

_____

_____

_____

_____

_____

_____

_____

_____

b. Describe how the frequency of an allele with a gene mutation might increase or decrease in a population over time.

_____

_____

_____

_____

_____

_____

_____

_____

_____

_____

_____

22. a. Construct an energy pyramid for a marine ecosystem with four levels. Label each trophic level of the energy pyramid, and provide an example of a marine organism at each level.

_____

_____

_____

_____

_____

_____

_____

_____

_____

_____

_____

b. Explain the difference between energy available at the bottom versus the top of the pyramid, and explain why this difference exists.

_____

_____

_____

_____

_____

_____

_____

_____

_____

_____

Chapter 36
Free-Response
Short Answers
Drill Answers and
Explanations

ANSWERS AND EXPLANATIONS

Free Response Short Answers Drill Answers and Explanations

1.   The driving force of evolution is natural selection or genetic variation.

   a.   Any three of the following—biogeography (study of the distribution of plans and animals in the environment), comparative anatomy (study of comparison of anatomy of different animals), embryology (study of organism development), molecular biology (study of nucleotides and amino acids), or paleontology (study of fossils)

   b.   The following are possible examples (others work as long as they are valid):

   Biogeography—similar traits in animals in geographically distinct areas such as the Galapagos Islands and mainland South America pointing to a common ancestor

   Comparative anatomy—homologous structures such as human arm, dog leg, and bird wing pointing to a common ancestor

   Embryology—all embryos look very similar early in development, all have gill slits

   Molecular biology—overlap in genetic code that increases when looking at increasingly similar species; chimpanzees and humans share 99% of genetic code

   Paleontology—evidence of great variety of organisms, evidence of natural selection (many organisms have died off and are no longer present today), and different lines of evolution (for example, primitive forms of humans leading to modern human)

2.   Natural selection occurs both internally by random mutation, and externally due to environmental pressures. Mutation is the only way for new versions of genes to appear in a population. For example, in a population of white moths, a random mutation to DNA could result in a moth that is black; this is an internal process. Usually, random mutations causing new traits will result in an organism that is less fit than those with the standard trait, and the individuals with the new trait will likely die off before they are able to reproduce. However, if the new trait produces some advantage instead of a disadvantage, the organism with the new trait will be more likely to survive and reproduce. For example, in an area with many factories producing soot that stains building and trees black, the black moth may have an advantage over the white moths (camouflage against the black backgrounds and less of a chance of being eaten). This is an example of an external force driving natural selection—changing environments favor certain traits.

3.   a.   A sex-linked gene is one that is carried on the X or Y chromosome; it is much more common for sex-linked genes to be found on the X chromosome than the Y chromosome, so most sex-linked genes are actually X-linked.

The easiest way to see that an abnormal trait was the result of a sex-linked gene is to notice a difference in the number of men versus the number of women who display that trait. Specifically, for X-linked genes, men will have the abnormal trait much more than women (because men have only one X chromosome while women have two X chromosomes).

   b.   Men will display the abnormal trait if they have a copy of an abnormal X-linked gene; because their sex chromosomes are XY, they only have one copy of the gene and thus will have the abnormal trait if they have the abnormal gene. If a woman has one copy of the abnormal X-linked gene, however, she will also have one normal copy (because her sex chromosomes are XX); so she will display the normal trait because the normal gene is dominant over the abnormal gene. A woman with one normal X-linked gene and one abnormal X-linked gene is called a "carrier". She will not display the abnormal trait, but if she passes the X chromosome with the abnormal gene onto her son, the son will be affected (the son will display the abnormal trait).

4.   a.   Fermentation is a process of anaerobic respiration, allowing cells to make ATP in the absence of any oxygen. Some anaerobic organisms that rely solely on fermentation include yeast cells and anaerobic bacteria. Pyruvic acid from glycolysis is combined with NADH. The byproducts of fermentation are ethanol (for yeast cells and some bacteria) or lactic acid (for other bacteria and muscle cells).

   b.   Human muscle cells can undergo fermentation when oxygen supply cannot keep up with the demand for ATP during exercise. While this allows for more ATP creation when there is otherwise not enough oxygen, it is very inefficient (only producing 2 ATPs), and also creates lactic acid (which is a toxin that causes much cramping).

5.   Viruses are nonliving things that are considered infectious. It is only by infecting a host cell and hijacking the host cell's machinery that they are able to perform functions unique to life such as protein synthesis and replication. Viruses are made up of just a protein capsid and within the capsid its genetic material (either DNA or RNA). In the viral lytic life cycle, a virus immediately begins to use host cell machinery for transcription and translation for more capsid proteins and makes copies of its genetic material. The virus then packages the genetic material into the capsid and lyses the cell, so many more copies of the virus are released and are able to go infect other cells. In the viral lysogenic life cycle, a virus first inserts its genetic material into the host cell's DNA and can remain there until something triggers it to enter into the lytic cycle.

6.   In an ecosystem, there are different levels of organisms in which energy flows from one level to the next. Producers, or autotrophs, have the most energy in an ecosystem and also make up the greatest biomass and the greatest number of organisms. Primary consumers, or heterotrophs who eat autotrophs (herbivores), have the next most energy in an ecosystem, and similarly, the next greatest biomass and number of organisms. This continues on up the food chain; secondary consumers have less energy, biomass, and numbers than primary consumers, and tertiary consumers have the least of all. As energy is transferred from one level to the next through the food chain, only about 10% of the total energy from one level makes it to the next level. The remaining 90% of the energy is lost as heat energy due to organism activities. Decomposers are not at a particular level on the food chain; they transcend the food chain because they will ultimately decompose all types of organisms (producers and all types of consumers).

7.   Polymerase chain reaction is a method to allow for fast replication and mass production of a specific sequence of DNA in the laboratory. The process is a modified version of physiological DNA replication, requiring the DNA of interest, RNA primers, a large supply of nucleotides (A's, T's, C's, and G's), and a special enzyme called *Taq* polymerase. All of these components are put into a tube that is then placed in a PCR machine, which thermocycles (meaning that the tube is heated, cooled, and then re-heated over and over again). Each time the tube is heated, the DNA is denatured, which means the hydrogen bonds between the two strands in the double-stranded DNA break, and the two strands separate. Then as the tube cools down, primers bind upstream of the DNA region of interest (primers are chosen based on the DNA sequence that you want to replicate). Then as it is warmed, *Taq* polymerase binds to the primers and adds nucleotides so that there is now one daughter strand annealed to each parental strand of DNA. The process is repeated the desired amount of times, and each time the amount of DNA in the test tube doubles. Some examples of why one might employ PCR are in crime scene investigations (to compare to a suspect's DNA), genetic testing (to look for disease mutations), tissue typing for organ transplantation, DNA sequencing, and phylogenic analysis of ancient DNA (for example from remains of human ancestors).

8. In the bacterial growth curve, the first flat part of the curve represents the "lag" phase, before the bacteria in the culture start to reproduce (1). The next part of the curve with the steep upslope represents the exponential growth, and because it eventually levels off, can be referred to as "log" or "exponential" phase (2). At some point, the bacterial colony will reach the carrying capacity for the environment and will level off. This part of the curve is called the "stationary" phase (3). Eventually, the bacteria will use up all of the resources available in the environment; when this happens, there will be nothing left to sustain the bacteria, and they will begin to die, so the number of bacteria will fall in the "death" phase (4).

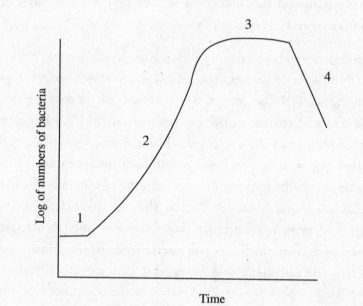

9. There are many examples of hormonal regulation via negative feedback in the body, and you could describe any of them to answer this question. Many of the hormones produced and released by the anterior pituitary provide good examples. For instance, the anterior pituitary releases adrenocorticotropic hormone (ACTH), which stimulates the adrenal cortex to release mineralocorticoids such as cortisol. When cortisol levels get high in the bloodstream, cortisol will bind to receptors on cells of the anterior pituitary, which inhibits secretion of ACTH. Then ACTH levels will drop, so the adrenal cortex will not be stimulated to produce cortisol, and cortisol levels will drop. Once cortisol levels are lower, there will not be an excess to bind the receptors on the anterior pituitary, so ACTH production will no longer be inhibited. Negative feedback like this is useful because it results in "hormonal homeostasis." This means that the levels of hormones will fall within a particular range that allows for optimal body function because if levels get too high, negative feedback will bring them back down, and once they get lower, the negative feedback will end so levels can begin to rise again.

10. Recombinant DNA is DNA that comes from two or more different source organisms; recombinant DNA technology allows us to transfer genes from one organism to another organism. A restriction enzyme that recognizes a short, specific DNA sequence is used to cut the DNA at recognition sequences at both ends, which isolates the gene of interest. The resulting double-stranded fragment has a sticky end that can hydrogen bond with the sticky ends of a specific type of plasmid called a cloning vector. Now the gene of interest is contained within a plasmid, called the recombinant DNA molecule. This recombinant DNA is then put into a bacterium by transformation, and the bacterium is grown in a culture to create bacterial clones that will also contain the plasmid. The recombinant DNA in the plasmid can be a gene of interest from a eukaryotic cell, which is then placed into a prokaryotic cell. The fact that DNA transcription and translation proceed in very similar ways in prokaryotes and eukaryotes allows the bacteria cell to transcribe and translate this eukaryotic gene; however, it is the special ability of prokaryotic cells to carry plasmids and transcribe and translate this extra-chromosomal DNA (outside of the bacterial chromosome) that allows the technology to work.

11. Skeletal muscle is found attached to bones and allows for voluntary body movement, while smooth muscle is found in the walls of the digestive tract and in blood vessels and is used for involuntary movement. Cardiac muscle is in the wall of the heart, and its movement is also involuntary. Skeletal muscle and cardiac muscle are striated, which allows for more organized shortening of muscles during contraction; smooth muscle is not striated. Only skeletal muscle is multinucleated. Cardiac muscle cells are joined by intercalated discs, which allows an impulse to travel quickly from one cell to the next for coordinated contraction. Skeletal muscle has the fastest speed of contraction, while smooth muscle contraction is the slowest.

12. Bacteria A at 37° C = (5 − 1.25)/120 = 0.03125 mL/min

Bacteria A at 5° C = (4 − 1)/120 = 0.025 mL/min

Bacteria B at 37° C = (2.25 − 1.75)/120 = 0.0042 mL/min

Bacteria B at 5° C = (0.2 − 0.1)/120 = 0.001 mL/min

Bacteria A seems to utilize aerobic metabolism more than Bacteria B, and the rate of aerobic metabolism for Bacteria A seems to be higher at higher temperatures. While Bacteria B seems to be utilizing aerobic metabolism at 37° C (to a lesser extent than Bacteria A), at 5° C, Bacteria B must be utilizing anaerobic metabolism because it has a very low rate of oxygen consumption at this lower temperature.

13. Your experiment should include two separate parts: one in which you change the pH for the reaction while keeping temperature, peroxidase concentration, and hydrogen peroxide (H_2O_2) concentration constant; the other in which you change the temperature while keeping pH, peroxidase concentration, and hydrogen peroxide (H_2O_2) concentration constant. You should mention the different pH levels and temperature points that you would use and also that you would measure the amount of water and/or oxygen present at different time points and plot this on a graph. Your hypothesis could be somewhat vague, such as "There is an optimal point for both pH and temperature at which the enzyme will work the best and thus the reaction will proceed the fastest" or you could predict at which temperature and pH level the enzyme will perform the best (it is not important that you predict the correct numbers because this is just a hypothesis, but you should have some rationale to back it up). You can also draw example graphs with potential data showing how you would identify the optimum pH and temperature for peroxidase.

14. The definition of a species includes the requirement that organisms of the same species must be capable of interbreeding to produce fertile offspring. Thus, one important piece of data to answer the question would be to see if two birds from the two geographically distinct regions are able to reproduce together and then if their offspring also are able to reproduce. Other types of data that could be used to answer the question would be sequencing the DNA of birds from the two areas to compare for similarities and differences, as well as closely comparing their morphological characteristics and/or the environments in which they live and their behaviors within that environment. It would be prudent to include in your answer that the ability to reproduce and have fertile offspring would be better evidence than the other types of observational data.

15. One major difference between prokaryotes and eukaryotes is that eukaryotes have membrane-bound organelles while prokaryotes do not; you can therefore name three of these organelles to answer this question. Be sure to then include how prokaryotes carry out similar functions without the given organelles. Some examples: eukaryotic cells have nuclei to house their chromosomal DNA, while prokaryotic circular DNA is found in the cytoplasm; mitochondria essential to aerobic respiration in eukaryotes (for Krebs cycle and electron transport chain/oxidative phosphorylation), while the cytoplasm and cell membrane serve similar roles in prokaryotes; the rough ER in eukaryotes is the site of protein synthesis/packaging/part of secretory pathway, while all proteins are synthesized in the cytosol and can occur at the same time as transcription in prokaryotes; chloroplasts in eukaryotes allow for light absorption and photosynthesis, while in prokaryotes cytosolic molecules and enzymes allow for these functions.

16. a. Water molecules are partially positively and negatively charged, making them polar, and they also have the ability to form hydrogen bonds.

b. These traits give rise to other special properties, including: cohesion (water molecules stick together), adhesion (water molecules stick to other substances), high specific heat capacity (requires a lot of heat energy to increase water's temperature), serving as a universal solvent (supports many different reactions), heat of vaporization (large energy requirement to vaporize water from liquid to gas), heat of fusion (large energy requirement to melt ice), and its thermal conductivity.

17. Normal development of organisms requires specific timing and coordination of events, which occurs via regulation of different genes. Differential expression of genes allows for different proteins to be present in different tissues and at different times within an organism. An important method for differential and sequential gene expression is the induction of transcription factors, which can turn genes "on" or "off" at different times and in different places. Homeotic genes are essential to timings and patterns during development. Soon after fertilization, embryonic induction is crucial to the correct timing of events during development. Later in life, the availability of different resources and differences in environment can induce different genes to be turned "on" or "off". This is essential as it would be a waste of resources for an organism to transcribe and translate genes for proteins that would not be useful at a certain time.

18. a. The positive slope part of the curve represents exponential growth due to a lack of limiting factors. Because there are plenty of resources so organisms quickly reproduce, and the growth rate greatly exceeds the death rate. In the flat part of the curve, density-dependent factors begin to play a role, as resources are decreased and waste products increase. This leads to stabilization and growth rate is approximately equal to the death rate.

    b. It would be reasonable to hypothesize that the addition of phosphate would cause another growth phase for the population, as this is a necessary resource for cellular energy and thus most cellular processes, including reproduction.

19. a. The cell cycle is necessary for growth, repair, and differentiation of cells. The order of the phases of the cell cycle is: Interphase (G1 → S → G2) → Mitosis (Prophase → Metaphase → Anaphase → Telophase). Interphase is a time for growth (in G1 and G2) and DNA replication (in S phase). Mitosis is when nuclear division occurs—in prophase, chromosomes condense, the nuclear envelope breaks down, and the spindle begins to assemble. In metaphase, the chromosomes align along the metaphasic plate. In anaphase, the sister chromatids are pulled to opposite sides of the cell. And in telophase, chromosomes disperse and the nuclear envelope is reformed. Cytokinesis is the process of cytoplasmic and cell content division, and it occurs at the same time as telophase.

    b. Regulation happens via checkpoints in the cell cycle, specifically at the end of G1 and G2 phases. Certain conditions must be met for the cell to progress past a checkpoint to the next step in the cell cycle. There are specific proteins involved in this regulation, such as MPF (maturation promoting factor) and CDKs (cyclin-dependent kinases), as well as hormones such as growth hormone. Mitosis is also inhibited in some cells via physical contact with other cells (indicating over-crowding).

20. Steroid hormones (such as estrogen, progesterone, and testosterone) are hydrophobic molecules which are able to cross the plasma membrane and bind to receptors within the cell (cytosolic receptors). The mechanism of action of steroid molecules is to alter gene expression by binding to the cytosolic receptor to change its conformation. This complex of the hormone and receptor then enters the nucleus and acts as a transcription factor, binding to a specific site on the DNA. The results of steroid hormone actions are slower to take effect than protein hormones but are sustained over a longer period of time.

Protein hormones, such as all hormones from the anterior pituitary (FSH, LH, ACTH, TSH, prolactin, and GH), are hydrophilic molecules that cannot cross the plasma membrane and instead bind to extracellular receptors. They typically act through second-messenger systems, such as G-protein receptors that go on to activate another enzyme. The results of these systems can alter gene expression or activate a specific pathway or enzyme system and usually occur quickly and temporarily in the target cell.

21. a. A single base-pair mutation occurs when a pair of bases (A, C, G, or T) is changed in the DNA; this can be a duplication, a deletion, or a substitution. When this occurs, the mutation in the DNA is then also included in the RNA that is transcribed, and then will be included in the protein via translation. If the single base-pair mutation results in a codon that codes for a different amino acid than the original codon, this will result in a difference in the protein. If this amino acid is in an important functional part of the protein, such as the active site, or an important structural site (such as disulfide bones, hydrogen bonds, or R-group interactions), this can result in a non-functional or less-functional protein or a changed function for the protein. A single-base pair mutation can also result in a premature stop codon, which would cause a truncated (shortened) protein, which is likely to result in an ineffective protein.

b. Hardy-Weinberg equilibrium occurs when certain stable population conditions are met, and it keeps a stable allele frequency in the population over time. The equation for allele frequency is given by: $p + q = 1$; the equation for genotype frequency is given by $p^2 + 2pq + q^2 = 1$. Natural selection would alter this allele frequency in a population, either decreasing the frequency of the mutant allele if it decreased fitness, or increasing the frequency of the mutant allele if it increased fitness. Differential fitness refers to reproductive success; if those with/without the mutant allele have more/less offspring, this will be reflected in the next generation's allele frequency. Other issues that may alter allele frequency include genetic drift due to small population size, non-random mating such as inbreeding, and migration.

22.   a.   Your pyramid should look something like this.

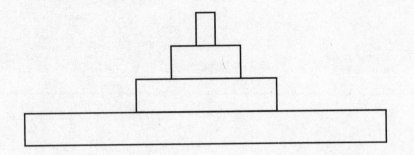

The labels should read (from bottom to top): Primary producer or autotroph → primary consumer or herbivore → secondary consumer or carnivore → tertiary consumer. If you mention decomposers, this should be written below the pyramid or off to the side, not as a part of the pyramid. Examples of marine organisms for each level include: algae → zooplankton → small fish → shark (other examples that fit in are fine).

   b.   There is more available energy at the bottom of the pyramid, and it decreases at each level as you move up. Energy is lost in the transfer from each level due to metabolic activities, heat, and other work requiring energy by the organisms.

Chapter 37
Free-Response Long
Answers Drill

FREE-RESPONSE LONG ANSWERS DRILL

1. Discuss the Hardy-Weinberg law.

   a. Explain what the Hardy-Weinberg law states.

   b. Describe the conditions that must be met for Hardy-Weinberg equilibrium to apply.

   c. Give an example of a situation in which Hardy-Weinberg equilibrium would not apply and explain why.

2. Explain the effect of an enzyme on a reaction.

   a. Describe what is and what is not changed by addition of an enzyme.

   b. Explain how temperature and pH can affect enzymes.

   c. Design an experiment to test how either a change in temperature or a change in pH affects a specific enzyme.

3. Explain the different parts of the *lac* operon (shown below) and how it functions as a whole.

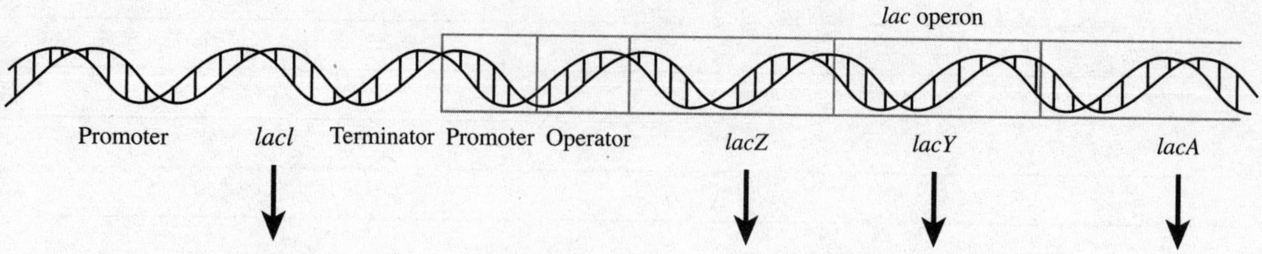
lac operon
Promoter
lacI
Terminator
Promoter
Operator
lacZ
lacY
lacA

4. Explain how humans have impacted the environment.

   a. Give specific examples of problems in the environment due to human actions.

   b. Discuss the extent of harm and the consequences stemming from these problems.

   _____

   _____

   _____

   _____

   _____

   _____

   _____

   _____

   _____

   _____

   _____

   _____

   _____

   _____

   _____

   _____

   _____

   _____

   _____

   _____

   _____

   _____

   _____

   _____

   _____

   _____

   _____

   _____

   _____

   _____

   _____

5. Information flow in cells is essential for the production of functional proteins that facilitate normal cellular activity.

  a. Explain the normal pathway for the flow of information in cells.

  b. Describe three different ways in which this flow of information can be regulated to allow for differences in protein synthesis at different times or in different cells.

  c. Information flow can also be changed by mutations. Describe three different types of mutations and how these would affect protein synthesis in a cell.

_____

_____

_____

_____

_____

_____

_____

_____

_____

_____

_____

_____

_____

_____

_____

_____

_____

_____

_____

_____

_____

_____

_____

_____

_____

_____

_____

_____

_____

_____

_____

_____

6. The small intestine is characterized by villi and microvilli. Villi are microscopic finger-like projections from the lining of the small intestine—each is made up of many cells (shown below). Microvilli are similar in shape but refer to projections on a single cell in the lining of the small intestine (also shown below).

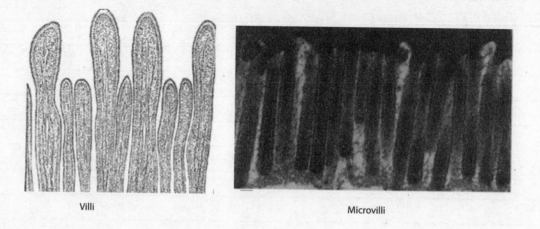

Villi   Microvilli

a. Explain the role that these two structures play within the small intestine.

b. Describe the underlying principle of cellular biology that relates to their function.

c. In certain human illness, such as Celiac disease, the villi are destroyed. Hypothesize how this might affect a person's digestion.

7. A never before seen species of rodent was discovered on a remote island. Different breeding crosses were performed, each using 50 males and 50 females, and the phenotypic results of the parental and offspring generations were recorded as shown below.

Cross 1—True-breeding long-tailed males were crossed with true-breeding short-tailed females, resulting in all long-tailed offspring in the F1 generation. F1 rodents were then crossed, with F2 results as shown in the table below.

F2 Phenotype	Male	Female
Long-tailed	1,860	1,900
Short-tailed	630	660

Cross 2—True-breeding normal-weight males were crossed with true-breeding overweight females, and all F1 generation offspring were overweight. F1 rodents were then crossed, with F2 results as shown in the table below.

F2 Phenotype	Male	Female
Normal-weight	580	660
Overweight	1,800	1,910

Cross 3—True-breeding long-tailed, overweight males were crossed with true-breeding short-tailed, normal-weight females, resulting in all F1 offspring being long-tailed and overweight. The F1 rodents were crossed with true-breeding short-tailed, normal-weight rodents, with the results shown in the table below.

Phenotype	Male	Female
Long-tailed, overweight	1,180	1,110
Long-tailed, normal-weight	110	150
Short-tailed, overweight	130	110
Short-tailed, normal-weight	1,120	1,090

a. What conclusions can be drawn from cross I results? Be sure to explain your reasoning for each conclusion.

b. What conclusions can be drawn from the results of cross II? Be sure to explain your reasoning for each conclusion.

c. What conclusions can be drawn from the results of cross III? Be sure to explain your reasoning for each conclusion.

Chapter 38
Free-Response
Long Answers
Drill Answers and
Explanations

ANSWERS AND EXPLANATIONS

Free-Response Long Answers Drill Answers and Explanations

1.   The Hardy-Weinberg law states that the relative frequencies of genotypes in a population remain constant over time. The dominant and recessive alleles are present in different proportions and these proportions stay consistent over time—one does not become increasingly common. The equation to describe allele frequencies is: $p + q = 1$ where p = dominant allele frequency and q = recessive allele frequency. The frequency of genotypes is given by the equation: $p^2 + 2pq + q^2 = 1$ where p^2 = homozygous dominants, pq = heterozygotes, and q^2 = homozygous recessives.

Conditions—1) large population, 2) no mutations, 3) no immigration or emigration, 4) random mating, and 5) no natural selection

Any example demonstrating one of the rules listed above being broken is acceptable. For example, if a small population of 50 crabs lived on the bank of a river, and a flood occurred which killed off 48 of the crabs, the new population would consist of only two individuals, and only their alleles would be represented in the population—the Hardy-Weinberg equilibrium would likely be violated given the small sample size that was greatly impacted by an environmental event. On the other hand, if a population of 5,000 crabs lived on the river bank and 96% of that population was destroyed in a flood, there would still be 200 crabs in the resulting population, and Hardy-Weinberg equilibrium would be more likely to be preserved.

2.   An enzyme increases the rate of reaction by lowering the activation energy required for the reaction to occur. An enzyme does not change the thermodynamics of a reaction—it will not cause a reaction that otherwise could not happen. Rather, an enzyme will speed up reactions that are able to occur. Most enzymes have both an optimal pH at which they operate and an optimal temperature at which they operate. For most enzymes in the body, their optimal temperature is body temperature (98.6° Fahrenheit). Optimal pH for body enzymes varies by their location and function—for example, some digestive enzymes function optimally in an acidic environment, which is provided by stomach acid.

Your experiment should specify if you are going to alter temperature or pH and how you intend to alter this and keep track of the results. For example, you could decide to look for the effect of temperature changes on a particular enzyme by placing reactants and enzyme (for an exothermic reaction) in three different settings—in the refrigerator, at room temperature, and on the stovetop. Then you could either set a time point at which you measure all three mixtures and determine the amount of product formed, or periodically test the three mixtures to see how long it takes to get to a certain amount of product. If you chose to measure the product level of all three mixtures after 5 minutes, you could then plot amount of product (y-axis) against temperature (x-axis) to see which is the enzymes optimal temperature to catalyze the reaction.

3.   The *lac* operon is found in some bacteria and is made up of four different parts—structural genes, the regulatory gene, the promoter gene, and the operator. There are three structural genes in the lac operon. As shown, the lacZ gene codes for beta-galactosidase, the lacY gene codes for lactose permease, and the lacA gene codes for thiogalactoside transacetylase. All three of these gene products are enzymes that are used for lactose digestion. The promoter is the place where RNA polymerase will bind to begin transcription. The operator is the region that controls if transcription of the structural genes will occur. The lacI gene is the regulatory gene, which codes for the repressor. The repressor is always transcribed and translated and will bind to the operator and block transcription of the structural genes. If lactose is present, however, it acts as an inducer and binds to the repressor, which causes it to fall off of the operator region, allowing transcription of the structural genes. Thus, a bacteria cell will only transcribe and translate the structural genes from the lac operon when lactose is present. This makes sense because it takes a lot of energy to translate mRNA into protein, so if lactose is not even available to the cell, it does not want to waste energy producing the enzymes that are needed for lactose breakdown. When lactose is present, it will bind to the repressor causing it to fall off, so the three enzymes for lactose breakdown will be produced and the cell will be able to breakdown lactose for energy.

4.   Humans have impacted the environment in many ways in the past several centuries, most of which have disturbed the pre-existing ecological balance of the environment. There are many potential examples of problems due to human activities. The greenhouse effect is due to increasing atmospheric concentrations of carbon dioxide from burning of fossil fuels and forests, leading to higher temperatures which may cause polar ice caps to melt and lead to flooding. The burning of fossil fuels has also created pollutants such as sulfur dioxide and nitrogen dioxide, which react with atmospheric water to produce acid rain. Acid rain lowers the pH of soil and aquatic ecosystems, which damages the organisms living in the ecosystem. Pollution has caused depletion of the ozone layer, which is problematic because ozone protects the surface of the earth from excessive ultraviolet radiation. Desertification is caused by overgrazing of land by animal farming and turns grasslands into deserts. Deforestation can lead to erosion and flooding. The extent of harm is far-reaching, and the effects are likely to continue to multiply as time goes on. In addition to flooding, global warming may lead to changes in precipitation, plant and animal populations, and agriculture. Depletion of the ozone layer may lead to increases in the incidence of cancer in humans and other organisms. Specific consequences of acid rain include changes in the pH of soil causing leaching of calcium, which will damage plant roots and stunt their growth and will also kill microorganisms in the soil which usually release nutrients that plants depend on. Acid rain can also kill newly hatched fish. Both desertification and deforestation, as well as pollution, distort or destroy the habitats of organisms, causing the extinction of animals and plants which inhabit these areas and an overall reduction in biodiversity.

5. The normal flow of information is as follows: DNA → RNA → protein. DNA → RNA is referred to as transcription and requires RNA polymerase enzymes. RNA → protein is called translation and requires ribosomes as well as tRNAs.

Some different methods of regulation of protein synthesis: RNA splicing (introns removed and remain in nucleus, exons spliced together and go into cytoplasm), repressor proteins (inhibit transcription or translation, silence genes/inactivate gene expression), transcription factors (promote or inhibit transcription, increase or decrease transcription of a specific DNA sequence), methylation (DNA or histone methylation inhibits transcription thereby inactivating specific DNA sequences, protects against restriction enzymes in prokaryotes), and siRNA (increases mRNA degradation thereby inhibiting translation).

Some common types of mutation: silent (single nucleotide change causing no change in amino acid sequence and fully functional protein), missense (single nucleotide change causing new codon and a different protein—may still function properly if not in a crucial position, e.g. active site of an enzyme), nonsense (single nucleotide change causing a stop codon and truncating the protein—likely devastating to protein function unless very close to the end of the amino acid sequence) or frameshift (nucleotide insertion or deletion changes the reading frame of the nucleotide sequence, leading to changes in the codon with the mutation and all subsequent codons, changing the amino acid sequence and the protein—likely cannot be a functional protein unless at the very end of the protein). Other possible mutations are at the chromosomal level—translocation (chromosome segment moves to a different place), nondisjunction (chromosomes do not separate during anaphase), duplication (chromosome segment doubles), deletion (chromosome segment deleted), inversion (reversal of a chromosome segment), or transposition (chromosome segment moves to a different site).

6. a. Both villi and microvilli serve to increase the absorptive surface area of the small intestine. This is important because the main role of the small intestine is digestion and absorption of nutrients. The increased surface area is useful because the nutrients enter cells via diffusion which is most effective over short distances—the increased surface area shortens the average distance traversed by nutrients and therefore allows for more effective diffusion of molecules into cells.

   b. The important underlying principle of biology is that surface-area to volume ratios affect biological systems' abilities to obtain resources (and eliminate wastes). Surface area is crucial to adequate exchange of materials, and in general, as cells increase in volume, they decrease in surface area. Special features such as villi serve to increase surface area.

   c. If villi are destroyed, nutrients will not be absorbed normally—there will be less effective abosorption of nutrients. This explains the common symptoms of diarrhea in celiac disease, as food moves too quickly through the digestive tract without being absorbed.

7. a. Cross I supports that tail-length is an autosomal trait that has classical dominance, with the long-tailed allele being dominant and the short-tailed allele being recessive. This is supported by the fact that all of the F1 generation, which are heterozygotes and had long-tails, and that the F2 generation had a ratio of 3:1 for long-tailed:short-tailed. It is not sex-linked because there was a very similar distribution of F2 phenotypes in males and females.

    b. Cross II supports that weight is also an autosomal trait that has classical dominance, with the overweight allele being dominant and the normal-weight allele being recessive. This is supported by the fact that all of the F1 generation, which are heterozygotes and overweight, and that the F2 generation had a ratio of 3:1 for overweight:normal-weight. It is not sex-linked because there was a very similar distribution of F2 phenotypes in males and females.

    c. Cross III supports that the two genes for tail-length and weight are linked because there is not the expected 1:1:1:1 ratio that would be expected for each of the four phenotypes listed in the table for this cross. The recombinant (new) phenotypes, long-tailed normal-weight and short-tailed overweight, are much less common than the phenotypes of the two rodents in the cross. These genes are 10 map units apart, calculated by the equation

$$\frac{\text{number of recombinants}}{\text{total number of offspring}} = \frac{500}{4,500} = 0.10$$

10% of the offspring are recombinants, which translates to 10 map units.

Part V
AP Biology
Practice Test

39 Practice Test
40 Practice Test Answers and Explanations

Chapter 39
AP Biology
Practice Test

AP® Biology Exam

SECTION I: Multiple-Choice Questions

DO NOT OPEN THIS BOOKLET UNTIL YOU ARE TOLD TO DO SO.

At a Glance

Total Time
1 hour and 30 minutes
Number of Questions
69
Percent of Total Grade
50%
Writing Instrument
Pencil required

Instructions

Section I of this examination contains 69 multiple-choice questions. These are broken into Part A (63 multiple-choice questions) and Part B (6 grid-in questions).

Indicate all of your answers to the multiple-choice questions on the answer sheet. No credit will be given for anything written in this exam booklet, but you may use the booklet for notes or scratch work. After you have decided which of the suggested answers is best, completely fill in the corresponding oval on the answer sheet. Give only one answer to each question. If you change an answer, be sure that the previous mark is erased completely. Here is a sample question and answer.

Sample Question

Sample Answer

Chicago is a
(A) state
(B) city
(C) country
(D) continent

Ⓐ ● Ⓒ Ⓓ

Use your time effectively, working as quickly as you can without losing accuracy. Do not spend too much time on any one question. Go on to other questions and come back to the ones you have not answered if you have time. It is not expected that everyone will know the answers to all the multiple-choice questions.

About Guessing

Many candidates wonder whether or not to guess the answers to questions about which they are not certain. Multiple-choice scores are based on the number of questions answered correctly. Points are not deducted for incorrect answers, and no points are awarded for unanswered questions. Because points are not deducted for incorrect answers, you are encouraged to answer all multiple-choice questions. On any questions you do not know the answer to, you should eliminate as many choices as you can, and then select the best answer among the remaining choices.

THIS PAGE INTENTIONALLY LEFT BLANK.

BIOLOGY

SECTION I

Time—1 hour and 30 minutes

Part A: Multiple-choice Questions (63 Questions)

<u>Directions:</u> Each of the questions or incomplete statements below is followed by four suggested answers or completions. Select the one that is best in each case and then fill in the corresponding oval on the answer sheet.

1. The stomach is a critical organ which performs all of the following functions EXCEPT

    (A) synthesizing pepsinogen to break down peptides into amino acids
    (B) mechanically breaking down of food by churning and mixing to produce chyme
    (C) generating HCl to denature peptides and destroy microbes
    (D) absorbing emulsified lipids and fats

Consider the following metabolic enzyme pathway:

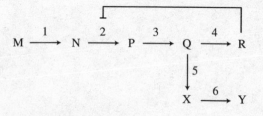

2. A sudden increase in substance R will have which of the following results?

    (A) An increase in substance P
    (B) A decrease in substance M
    (C) Increased activity of enzyme 3
    (D) Decreased activity of enzyme 4

3. During an action potential, the membrane potential reaches a maximum at +35 mV. Which of the following occur at this voltage?

    I. Na^+/K^+ pumps close
    II. Voltage-gated Na^+ channels open
    III. Voltage-gated K^+ channels open

    (A) I only
    (B) III only
    (C) II and III only
    (D) I, II, and III

4. Which of the following best describes the role of the corpus luteum during the menstrual cycle?

    (A) It produces follicle-stimulating hormone (FSH) and luteinizing hormone (LH), which stimulate follicle development and release.
    (B) It produces estrogen, which triggers menstruation approximately 14 days after ovulation.
    (C) It produces progesterone, which promotes growth of glands and blood vessels in the endometrium following ovulation.
    (D) It produces testosterone, which triggers ovulation after a surge in LH production.

GO ON TO THE NEXT PAGE.

Consider the following appendages of a human arm and a dog's front leg:

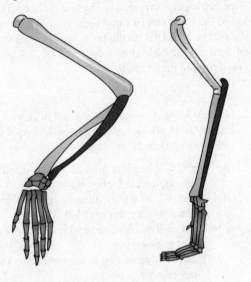

5. These homologous structures reflect a common ancestry despite clear differences in function. Which of the following terms best describes the events that gave rise to these structures?

(A) Divergent evolution
(B) Convergent evolution
(C) Sympatric speciation
(D) Allopatric speciation

6. All of the following describe differences in translation and trafficking events for secreted and cytosolic proteins EXCEPT

(A) cytosolic genes are normally translated by free ribosomes, whereas secreted genes are normally translated by ER-associated ribosomes
(B) secreted genes often contain ER-targeting signal peptides, whereas cytosolic genes do not
(C) cytosolic mRNA strands begin with the AUG start codon, whereas secreted mRNA strands begin with the UGA start codon
(D) secreted proteins are trafficked through the secretory pathway, whereas cytosolic genes are not

7. Which of the following is NOT an assumption made with a population in Hardy-Weinberg equilibrium?

(A) There will be no immigration or emigration into or out of the population.
(B) There will be a sufficiently large population to limit allelic difference due to random chance.
(C) There will be no mutations, which arise in the population.
(D) There will be natural selection, which drives allelic variability.

8. Abandoned birds have been hatched under the wings of a glider aircraft and have learned flight and traditional migratory routes by following the aircraft. This represents an example of which of the following types of learning?

(A) Imprinting
(B) Habituation
(C) Operant conditioning
(D) Insight

9. Synapsis and crossing-over are critical events, which result in genetic diversity of the cell. Which step of meiosis does synapsis and crossing-over events occur?

(A) Metaphase I
(B) Metaphase II
(C) Prophase I
(D) Prophase II

GO ON TO THE NEXT PAGE.

Questions 10-14

Pulse oximeters measure the amount of arterial oxygen saturation by evaluating transmission of red and infrared light through tissues such as a finger. Because deoxygenated hemoglobin absorbs more red light and less infrared light than oxygenated, a general ratio may be computed. An experiment was performed to evaluate the role of blood pH on the binding of oxygen to hemoglobin in mice. A pulse oximeter was applied and the blood pH was changed with an adjusted saline solution to 7.2, 7.4, or 7.6. The oxygen saturation was determined as the partial pressure of oxygen was increased in the graph below.

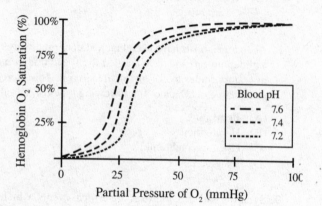

12. How does the absorption of red and infrared light at higher blood pH values differ from lower pH values by pulse oximetry in 50 mmHg O_2?

(A) There will be higher red and lower infrared absorptions at higher blood pH.
(B) There will be higher infrared and lower red absorptions at higher blood pH.
(C) There will be equal red and infrared absorptions at higher blood pH.
(D) The results cannot be determined from the information provided.

13. In which of the following cardiovascular structures would the lowest oxygen saturation be detected?

(A) Aorta
(B) Pulmonary artery
(C) Pulmonary vein
(D) Capillary beds in the finger

10. Which of the following best describes the relationship between blood pH and hemoglobin oxygen saturation in mice?

(A) As blood becomes more acidic, the oxygen saturation in hemoglobin increases at the same partial pressure of oxygen.
(B) As blood becomes more acidic, the oxygen saturation in hemoglobin decreases at the same partial pressure of oxygen.
(C) As blood becomes more acidic, the oxygen saturation in hemoglobin remains the same until a partial pressure above 75 mmHg.
(D) There is no relationship between blood pH and hemoglobin oxygen saturation.

14. Erythrocytes (red blood cells) carry high concentrations of hemoglobin. Where are erythrocytes made?

(A) Liver
(B) Spleen
(C) Kidney
(D) Bone marrow

11. How does the oxygen binding of fetal hemoglobin compare to adult hemoglobin?

(A) Fetal hemoglobin has higher binding affinity.
(B) Fetal hemoglobin has lower binding affinity.
(C) Fetal hemoglobin has equal binding affinity.
(D) There is no difference in the structures of fetal and adult hemoglobin.

GO ON TO THE NEXT PAGE.

15. Which of the following types of genetic mutations will be least likely to alter the length of the resulting protein?

    (A) Missense mutation
    (B) Nonsense mutation
    (C) Insertion
    (D) Duplication

16. During photosynthesis, what is the purpose of the light-independent reactions?

    (A) The light-independent reactions generate NADPH and ATP to be used by the light dependent reactions to produce sugar.
    (B) The light-independent reactions generate chlorophyll to be used in the light dependent reactions to produce sugar.
    (C) The light-independent reactions use NADPH, ATP, and CO_2, to generate sugar.
    (D) The light-independent reactions convert sugar into cellulose for storage until needed.

17. An experiment was performed crossing different species of pea plants. In one cross between two heterozygotes of the same species, for pea color the ratio of dominant to recessive plants for color was much less than the expected 9:3:3:1 ratio. Assuming pea color is controlled by two genes, which of the following best explains this result?

    (A) The deviation in phenotypic ratio for pea color was due to incomplete dominance.
    (B) The deviation in phenotypic ratio for pea color was due to codominance.
    (C) The deviation in phenotypic ratio for pea color was due to linkage.
    (D) The deviation in phenotypic ratio for pea color was due to epistasis.

GO ON TO THE NEXT PAGE.

Questions 18-19

Four new species of archaea named A – D, were recently discovered in the thermal pools in Yellowstone National Park. The nucleotide sequences were determined for the bacterial rRNA and the table below reflects differences between species.

Nucleotide Differences

Species	A	B	C	D
A	0	16	5	6
B		0	19	21
C			0	2
D				0

19. Which of the phylogenetic trees below is most consistent with the archaea rRNA data provided?

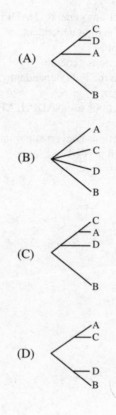

(A)

(B)

(C)

(D)

18. What is the role of rRNA in the cell?

(A) rRNA encodes the messages for genes which are then translated by ribosomes in the cytoplasm.

(B) rRNA anticodons are paired with codons by ribosomes to assemble nascent peptides during translation.

(C) rRNA is the nucleic acid component of ribosomes and binds RNA messages during translation.

(D) rRNA encodes a gene in the nucleus of a cell prior to transcription by host polymerases.

GO ON TO THE NEXT PAGE.

20. Shown below is an enzyme that is inhibited by binding of inhibitor Y. Which of the following accurately describes the function of this inhibitor?

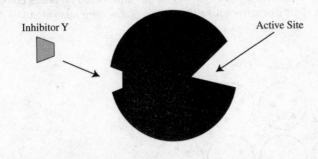

Inhibitor Y

Active Site

(A) Inhibitor Y competes with substrate for access to the active site.
(B) Inhibitor Y binds an allosteric site causing a conformational change to block the active site.
(C) Inhibitor Y binds the substrate preventing its subsequent binding to the active site.
(D) Inhibitor Y blocks translation of the enzyme.

21. What is the order from largest to smallest of structures containing chlorophyll in a plant cell?

(A) Mitochondria, Chloroplasts, Grana, Thylakoids
(B) Chloroplasts, Grana, Stoma, Thylakoids
(C) Chloroplasts, Grana, Thylakoids, Antenna Complexes
(D) Chloroplasts, Thylakoids, Grana, Stoma

22 A bacterial cell is treated with radiolabeled sulfur-35 (^{35}S) methionine and cysteine, if bacteriophages infect the cell. What component, if any, of the progeny bacteriophage structure will contain the most ^{35}S?

(A) Bacteriophage Genome
(B) Capsid (Bacteriophage Protein Coat)
(C) Envelope (Bacteriophage Membrane)
(D) Bacteriophage Nuclear Envelope

23. The DNA sequence of a gene begins with the following:

TAC CCC ATC GGC CCT

Which of the following would be the sequence of transcribed mRNA from the DNA segment shown above?

(A) ATG GGG TAG CCG GGA
(B) GCA AAA CGA TTA AAG
(C) AUG GGG UAG CCG GGA
(D) AGG GCC GAU GGG GUA

24. Over the last fifty years, an abrupt decrease in average temperature on an island has resulted in a gradual change from wind-blown forests of short conifers to open fields containing permafrost and few trees. Which of the following transitions in the biome is occurring?

(A) Taiga to deciduous forest
(B) Taiga to tundra
(C) Tundra to taiga
(D) Tundra to grasslands

25. Sentry meerkats stand on alert to detect predators and will continue to bark if they spot a predator to warn fellow meerkats to hide despite often drawing the attention of the predator to themselves. This is an example of which of the following types of social behaviors?

(A) Agonistic behavior
(B) Dominant hierarchy
(C) Territoriality
(D) Altruistic behavior

GO ON TO THE NEXT PAGE.

Questions 26-30

New hospital strains of *E. coli* (named A, B, and C) are being evaluated for sensitivity to antibiotics. The bacteria have been spread on plates to isolate pure cultures. The pure colonies were selected, suspended in growth media, and spread on plates containing antibiotics ampicillin (Amp⁺), kanamycin (Kan⁺), penicillin (Pen⁺), or tetracycline (Tet⁺). The plates were incubated at 37°C for 24 hours. The experimental setup and results are shown below.

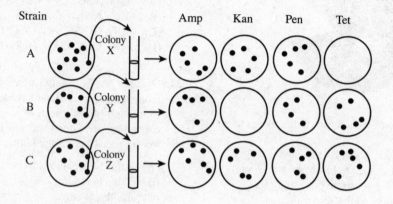

26. Based on the data shown, which strain is the least sensitive to the antibiotics tested?

    (A) Strain A
    (B) Strain B
    (C) Strain C
    (D) They are all equally sensitive.

27. A patient has an infection with *E. coli* strain B. Which of the antibiotics should be prescribed?

    (A) Ampicillin
    (B) Kanamycin
    (C) Penicillin
    (D) Tetracycline

28. Suppose a follow-up experiment is performed which shows that the *E. coli* strains are practically genetically identical and only very in their expression of antibiotic resistance genes. All of the following hypotheses are supported by these new findings EXCEPT

    (A) *E. coli* strains A, B, and C are derived from a single common bacterial strain.
    (B) *E. coli* strains A, B, and C have acquired different plasmids, small pieces of DNA, which confer resistance to specific antibiotics.
    (C) *E. coli* strain C has a unique flagellum and attachment proteins, which has provided an advantage in acquiring resistance genes.
    (D) *E. coli* strain C evolved from strain B.

29. Which of the following cellular structures would you NOT expect to find in *E. coli*?

    (A) Ribosomes
    (B) Plasma membrane
    (C) Cell wall
    (D) Nucleus

30. Assuming there are no bacteriophages present, the process by which the bacteria likely acquired the resistance genes from their environment is called

    (A) transduction
    (B) transfection
    (C) transformation
    (D) PCR

GO ON TO THE NEXT PAGE.

31. A catalyst has been added to an exergonic reaction. What impact will the catalyst have on the energetics of the reaction?

    (A) The catalyst will reduce the energy of the products making the reaction more spontaneous.
    (B) The catalyst will reduce the energy of the reactants making the reaction more spontaneous.
    (C) The catalyst will lower the activation energy making the reaction more efficiently proceed to completion.
    (D) The catalyst will raise the activation energy making the reaction more efficiently proceed to completion.

32. The pancreas secretes a variety of digestive enzymes into the small intestine. Trypsin and chymotrypsin are secreted by the pancreas and are most similar to which of the following?

    (A) Lipase
    (B) Pepsin
    (C) Amylase
    (D) Lactase

33. Which of the following immune cells is responsible for producing antibodies?

    (A) Phagocytes
    (B) B cells
    (C) Cytotoxic T cells
    (D) Helper T cells

34. Which of the following organs of the digestive system is considered an accessory organ, meaning that it is not part of the GI tract?

    (A) Stomach
    (B) Small intestine
    (C) Liver
    (D) Esophagus

35. Global warming has caused the size of some species of cacti to decrease. The change in cactus size has triggered a decrease in the size of cactus wrens, which nest in holes in cacti. This is an example of which of the following types of selection?

    (A) Sexual selection
    (B) Stabilizing selection
    (C) Disruptive selection
    (D) Directional selection

36. In a community, prairie dogs eat grasses, brush, and flowering plants and are in turn eaten by foxes and coyotes? Which of the following accurately describes the niche of prairie dogs in this community?

    (A) Producers
    (B) Primary consumers
    (C) Secondary consumers
    (D) Decomposers

GO ON TO THE NEXT PAGE.

37. The nervous system will develop from which of the following germ layers during organogenesis?

    (A) Mesoderm
    (B) Endoderm
    (C) Zygoderm
    (D) Ectoderm

38. Which of the following are similarities between smooth muscle and cardiac muscle cells?

    I. They have a single nucleus.
    II. They are under involuntary control.
    III. They are striated.

    (A) I only
    (B) I and II only
    (C) I and III only
    (D) I, II, and III

39. Robert Wadlow had gigantism and remains the tallest human being in recorded history at a final height of 8 feet 11 inches tall. Scientists attribute his height due to excessive production of growth hormone, produced by which of the following?

    (A) Hypothalamus
    (B) Adrenal gland
    (C) Pituitary gland
    (D) Thyroid gland

GO ON TO THE NEXT PAGE.

Questions 40-43:

Scientists evaluating inhibitors of mitosis and have identified a chemical called CIS, which results in all mitotic cells being arrested at the stage shown below.

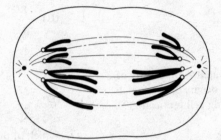

40. During which stage of mitosis are the cells arrested?

   (A) Prophase
   (B) Metaphase
   (C) Anaphase
   (D) Telophase

41. Which of the following is a possible explanation for how the inhibitor is arresting the cells?

   (A) The inhibitor is blocking the complete contraction of the microtubules.
   (B) The inhibitor is preventing the separation of the chromatids at the centromere.
   (C) The inhibitor is blocking the dissolution of the nuclear membrane.
   (D) The inhibitor is blocking the formation of mitotic spindles.

42. The cell shown above may be any of the following types of cells EXCEPT

   (A) spermatocyte
   (B) hepatocyte (liver cell)
   (C) lymphocyte (white blood cell)
   (D) epithelial cell

43. Which of the following events is expected to happen next after the inhibitor degrades?

   (A) The chromosomes will align in the center of the cell.
   (B) The nuclear envelope will reappear and the cell will begin cytokinesis.
   (C) Synapsis and crossing-over will occur.
   (D) Complete copies of the cells' genetic material will be made in preparation for cell division.

GO ON TO THE NEXT PAGE.

44. A missense mutation has resulted in a change from a glycine to a proline. This change is limiting development of a critical alpha helix due to increased restriction in the flexibility of the polypeptide backbone. Which of the following levels of protein structure has been most affected?

    (A) Primary
    (B) Secondary
    (C) Tertiary
    (D) Quaternary

45. Down syndrome is a condition associated with trisomy of chromosome 21. Trisomy most often occurs due to which of the following types of defects during cell replication?

    (A) Translocation
    (B) Nondisjunction
    (C) Duplication
    (D) Deletion

46. A baby boy has been born to a man with an A blood type and a woman with an O blood type. Assuming the child has an older sibling with an O blood type, what is the probability that the he will have an A blood type?

    (A) 0

    (B) $\dfrac{1}{4}$

    (C) $\dfrac{1}{2}$

    (D) $\dfrac{2}{3}$

47. Which of the following structures of the respiratory tract is the primary location of gas exchange?

    (A) Terminal bronchioles
    (B) Bronchi
    (C) Alveoli
    (D) Tracheae

GO ON TO THE NEXT PAGE.

Questions 48-51: Refer to the image below of a nephron to answer the following questions.

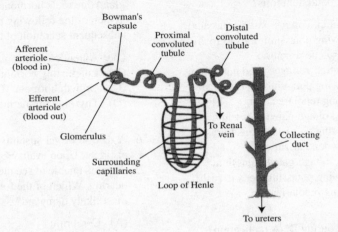

48. Aldosterone plays a critical role in regulating blood pressure through absorption of extra sodium from the filtrate into the blood. Which of the following structures of the nephron does aldosterone target?

(A) Bowman's capsule
(B) Proximal convoluted tubule
(C) Loop of Henle
(D) Distal convoluted tubule

49. Coffee is an antidiuretic hormone (ADH) inhibitor and acts on the collecting duct of the nephron. Which of the following directly occurs due to the ADH inhibition activity of coffee?

(A) Coffee increases water reabsorption leading to increased urine production.
(B) Coffee decreases water reabsorption leading to increased urine production.
(C) Coffee increases sodium reabsorption leading to increased urine production.
(D) Coffee decreases sodium reabsorption leading to increased urine production.

50. Which of the following should NOT be found in the filtrate in Bowman's capsule?

(A) Salt
(B) Water
(C) Urea
(D) Albumin

51. Many organisms produce urea as a waste byproduct. Where does urea come from?

(A) Urea is a byproduct of lipid catabolism.
(B) Urea is a byproduct of nucleic acid catabolism.
(C) Urea is a byproduct of protein catabolism.
(D) Urea is a byproduct of carbohydrate catabolism.

GO ON TO THE NEXT PAGE.

52. During an action potential, which of the following best describes the movement of potassium ions?

    (A) At the threshold potential, voltage-gated potassium channels open allowing potassium ions to rush into the cell causing depolarization.
    (B) At the threshold potential, voltage-gated potassium channels open allowing potassium ions to rush out of the cell causing repolarization.
    (C) At the peak potential, voltage-gated potassium channels open allowing potassium ions to rush into the cell causing depolarization.
    (D) At the peak potential, voltage-gated potassium channels open allowing potassium ions to rush out of the cell causing repolarization.

53. What is the role of helicase during DNA replication?

    (A) Helicase adds nucleotides to existing strands creating a new helix.
    (B) Helicase unwinds the double helix into two strands.
    (C) Helicase ligates together the Okazaki fragments in the lagging strand.
    (D) Helicase catalyzes the synthesis of RNA primers for polymerization.

54. Chlorofluorocarbons (CFCs) are being phased out of use in aerosol cans following the Montreal Protocol to reduce their influence on which of the following?

    (A) Greenhouse effect
    (B) Ozone depletion
    (C) Acid rain
    (D) Deforestation

55. The pituitary gland is commonly referred to as the master gland due to its hormonal influence on other glands. Which of the following pituitary hormones results in subsequent secretion of thyroxine?

    (A) Adrenocorticotropic hormone (ACTH)
    (B) Luteinizing hormone (LH)
    (C) Growth hormone (GH)
    (D) Thyroid-stimulating hormone (TSH)

56. A racecar driver sustains brain damage following a car accident. Upon evaluation in the emergency room, the driver is unable to see clearly and is having difficulty hearing. Which of the following parts of the brain was most likely damaged?

    (A) Cerebrum
    (B) Cerebellum
    (C) Pons
    (D) Hypothalamus

57. A plant has three independently assorting traits AaBbCcDd. What fraction of the gametes are expected to have the genes abcd?

    (A) 0

    (B) $\frac{1}{2}$

    (C) $\frac{3}{8}$

    (D) $\frac{1}{16}$

GO ON TO THE NEXT PAGE.

Questions 58-59: Refer to the image below of ecological succession to answer the following questions.

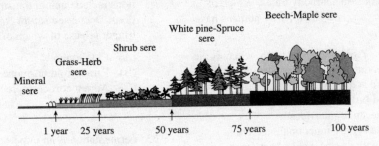

58. A forest is dominated mostly by confers with a small number of young deciduous trees? Based on the scale above, approximately how old is the forest?

   (A) 15 years old
   (B) 35 years old
   (C) 65 years old
   (D) 90 years old

59. Suppose a forest fire hits an old forest of mostly beech and maple trees? How will the pace of secondary succession compare to the initial ecological succession in the area?

   (A) The growth of the secondary community will be faster than the initial ecological succession.
   (B) The growth of the secondary community will be slower than the initial ecological succession.
   (C) The growth of the secondary community will be about the same as the initial ecological succession.
   (D) The growth of the secondary community will depend on whether lichens survived the forest fire.

GO ON TO THE NEXT PAGE.

60. Salivary amylase and pancreatic amylase perform very similar functions in digesting carbohydrates. Which of the following provides an rationale for why humans have both enzymes?

    (A) Salivary amylase rapidly digests starch, whereas the pancreatic amylase rapidly digests cellulose.
    (B) Salivary amylase is inactivated in the stomach due to low pH and further carbohydrate digestion is required in the small intestine.
    (C) Salivary amylase is only functionally active in combination with the mechanical action of chewing, whereas pancreatic amylase is much more versatile.
    (D) Salivary amylase and pancreatic amylase are remnants of evolution from other primates.

61. An animal cell is bathed in a hypertonic solution. Which of the following is expected to occur?

    (A) The cell will remain the same size due to equivalent movement of water into and out of the cell.
    (B) The cell will shrink due to movement of water out of the cell.
    (C) The cell will expand due to movement of water into the cell.
    (D) The cell will stay the same size due to its cell wall, but its membranes will shrink.

62. Calcium is critical for a wide variety of physiologic functions include neural transmission and muscle contractions. Decreased serum concentrations of calcium will trigger release of which of the following hormones?

    (A) Calcitonin
    (B) Parathyroid hormone (PTH)
    (C) Adrenocorticotropic hormone (ACTH)
    (D) Oxytocin

63. Fermentation is an important process for eukaryotes. However, the end products for organisms often differ. What are the end products of human and yeast fermentation, respectively?

    (A) Lactic acid and methanol
    (B) Lactic acid and ethanol
    (C) Ethanol and pyruvate
    (D) Ethanol and glutaraldehyde

GO ON TO THE NEXT PAGE.

Part B: Grid-in Questions (6 Questions)

<u>Directions:</u> This section consists of questions that require numeric answers. Calculate the correct answer to each question.

64. In pea plants, purple flowers (F) are dominant over white flowers (f), and green pods (P) are dominant over yellow pods (p). Suppose a student performed a cross between a purple-flowered pea plant with green pods that is heterozygous for both traits and a yellow-flowered pea plant with green pods that is heterozygous for pod color. Assuming independent assortment, what fraction of progeny pea plants will have purple flowers and green pods? Give your answer in fraction form.

65. Earwax moisture is a human trait controlled by a single gene. Wet earwax is dominant over dry earwax. Assume that 4% of a population has dry earwax and that the population is in Hardy-Weinberg equilibrium. What is the frequency of the dominant allele in the population? Provide your answer to the nearest tenth.

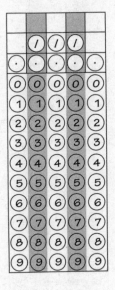

GO ON TO THE NEXT PAGE.

66. The world population has grown steadily over the last
    800 years despite as shown in the figure below. In 1450,
    the black plague pandemic struck Europe killing approxi-
    mately 1 in 4 people. What was the world population at
    the time of the black plague. Give your answer in mil-
    lions of people to the nearest tenth place.

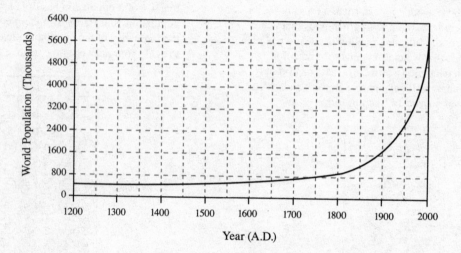

GO ON TO THE NEXT PAGE.

67. Shown below is the carbon flow for an aquatic food web. What is the carbon (in g/m^2) released by respiration of plankton. Provide your answer to the nearest whole number.

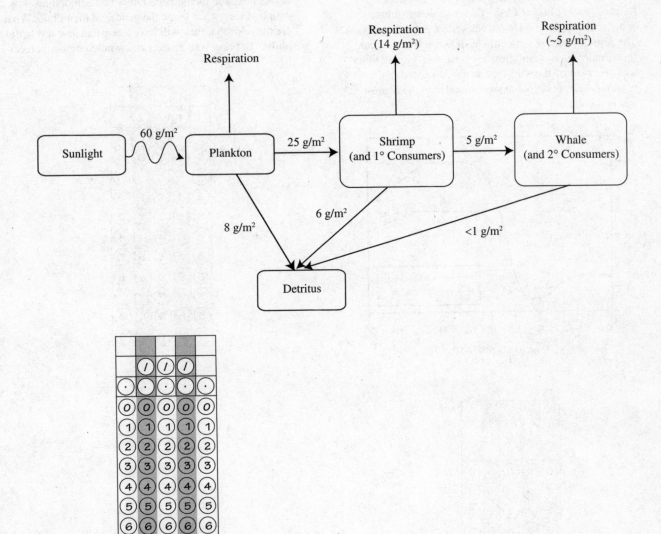

GO ON TO THE NEXT PAGE.

68. A graduate student performs an experiment comparing a wild-type strain and a temperature-sensitive (*ts*) mutant strain of a virus to evaluate the defect in replication of the virus at elevated temperature. In the experiment, the student inoculated 25 cm² flasks with 20 plaque-forming units (PFU) of virus. The virus was incubated at 40°C and took samples for virus titers every 2 hours. The results of the experiment are shown below. At peak (maximum) titers, approximately what fold greater titers was observed by the wild-type strain compared to the *ts* mutant? Express your answer rounded to the nearest whole number.

69. Hemophilia is a sex-linked disorder characterized by the inability to properly form clotting. A couple discuss with a family geneticist about the chances that their unborn child will inherit hemophilia. The child's father has hemophilia and its mother's father had hemophilia. The couple does not yet know the gender of their child. What are the odds that they will have a boy that also has hemophilia? Express your answer as a whole number percentage.

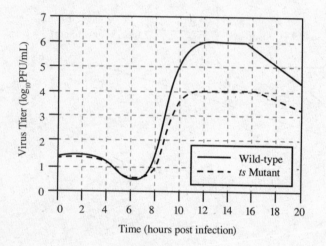

END OF SECTION I

BIOLOGY
SECTION II
Time – 1 hour and 30 minutes

Directions: Questions 1 and 2 are long free-response questions, which should require approximately 20 minutes each to answer. Questions 3 – 8 are short free-response questions, which should require approximately 6 minutes to complete each. Read through each question or prompt carefully and write your complete response. It is important to read carefully before you begin writing.

1. A biologist has been studying a family of wolves, which are afflicted with an apparent genetic disease. Below is a pedigree exhibiting the occurrence of the disease through three generations. Ecological surveys have shown that the disease is generally quite rare in wolves and no wolves, which have joined the pack showed any sign of the disease.

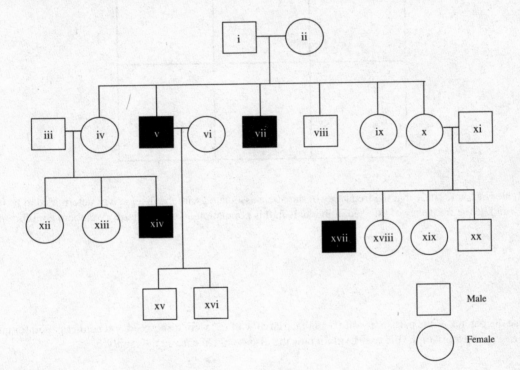

a. Using the data provided in the pedigree above, **explain** what type of genetic disease (e.g. autosomal, sex-linked) is afflicting the family of wolves and whether or not the disease is domain. **Justify** your answer.

b. If wolf xvii breeds with xii, **calculate** what is the likelihood that their first female pup will be afflicted with the disease.

GO ON TO THE NEXT PAGE.

c. **Draw** a Punnett square using the box provided below to display the cross between wolves v and vi. Be sure to include the genotypes of the parents in your cross. **Explain** using your results why neither male pup of the wolves has the disease. Include in your answer what you would expect to see, if the wolves have a female pup.

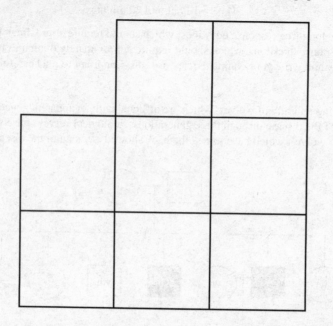

d. In a greater genomic study that the frequency of the allele associated with the disease was determined to be 0.1. **Calculate** what would be the frequency of affected individuals in this population assuming Hardy-Weinberg equilibrium?

e. Suppose the parents of the patriarch wolf (i) and matriarch wolf (ii) were discovered and neither parent for either wolf had the disease or a history of the disease. **Explain** how the disease likely entered the family.

GO ON TO THE NEXT PAGE.

2. A scientist has recently discovered a new DNA ligase in a species of thermophilic bacteria. The DNA ligase performs similar functions to other eukaryotic ligases; however, the scientist quickly hypothesizes that the ligase may be more resistant to degradation at higher temperatures than most ligases due to the optimal growth temperature of the bacteria. To evaluate the activity of the new DNA ligase, the scientist determines the relative activity of enzyme over changes in temperature as shown below.

LIGASE ACTIVITY VERSUS TEMPERATURE

Temperature (°C)	4	25	37	45	60	75	90
Relative Activity (%)	0	5	18	40	70	97	30

a. **Draw** a graph using the data above and provided axes, which show the impact of temperature on the change in the relative activity of the recently discovered DNA ligase. **Explain** the shape of the curve between 75° to 90°.

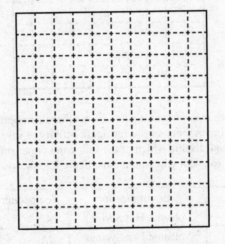

b. A follow-up experiment is performed to determine the maximum activity of the enzyme toward a fixed amount of substrate at 72° C. **Explain** using the curve below why increases in enzyme concentration above 0.5 µM have no effect on enzyme activity.

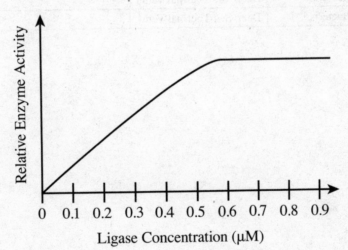

GO ON TO THE NEXT PAGE.

c. Assume that the hypothesis that DNA ligase activity is optimized for the optimal temperature of the organism in which it grows is correct. **Draw** a graph that displays what you believe the relative activity of a DNA ligase isolated from an enteric (gut) strain of *E. coli* would like over changes in temperature. **Justify** your drawing based on the hypothesis provided.

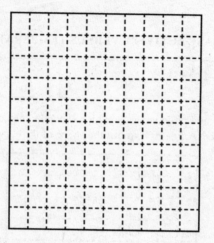

d. Although the discovery of the ligase is important, ligases from other thermophilic organisms have been recovered. The data shown below displays the optimum temperature for ligase activity of several different types of organisms and the environments which they were found. **Define** which of the following organisms share similar ligase characteristics to the one described above and what environment you would most likely find the yeast growing. **Justify** your answers.

Organism	Environment	Temperature for Optimum Ligase Activity (°C)
Thermobacillus Archaea	Volcanic Hot Springs	92
Hyperthermococcus Archaea	Volcanic Hot Springs	88
Thermoaquaticus Bacteria	Volcanic Hot Springs	83
Pyrospirallis Bacteria	Deep Sea Thermal Vents	76
Thermoacidophilus Fungi	Deep Sea Thermal Vents	72
Tyropyrococcus Bacteria	Deep Sea Thermal Vents	56

GO ON TO THE NEXT PAGE.

3. The kakapo (*Strigops habroptilus*) is a flightless, ground-dwelling parrot which used to roam the undergrowth of forests in New Zealand. However, introduction of cats by European colonization of the islands has resulted in a severe collapse of the population of kakapo. The last remaining native populations were discovered in the forested mountainous region in the Southern Alps of the south island. Interestingly, the leaping and gliding ability of the kakapo, which is an excellent climber, has increased in steadily over the last 100 years.

   a. **Propose** a hypothesis for why the kakapo has recently seen an increase in leaping and gliding ability.

   b. **Explain** an experiment for how you would test the hypothesis written in part (a) and how the data acquired would help you address your hypothesis.

4. A population of *E. coli* is being grown from a single colony selected on a plate in a 1 L flask in 250 mL of LB broth (growth media) at 37° C. The population growth is shown below.

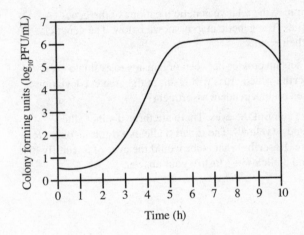

   Write a paragraph where you **explain** the biological events occurring that defines the shape of the curves between a) 2 and 5 hours and b) after 9 hours.

GO ON TO THE NEXT PAGE.

5. A unique population of algae has been discovered growing at a depth of 30 meters below the surface of the ocean. Two distinct populations of algae which appear to have similar structural and growth characteristics to the newly discovered species have been identified at a shallow depth of only 10 meters below the surface.

   a. **Describe** TWO different types of data that you could collect that would answer the question: is the deep-growing population of algae discovered related to the two populations growing at 10 meters?

   b. **Explain** how the data acquired in part (a) would help you directly answer the question.

6. A geneticist has recently determined the relative genetic locations of the genes (A – D) controlling eye color for fruit flies. The genetic map is shown below. The geneticist begins to setup crosses to test their results.

   a. Suppose that the geneticist sets up crosses between organisms to evaluate every possible pair of genes. **Describe** which cross will result in the greatest deviation from the expected ratio based on independent assortment?

   b. Gene A encodes for the base color of the eyes. There are three distinct allele variants: α (red), β (blue), and γ (yellow). The α and β alleles exhibit incomplete dominance and γ is recessive. **Describe** what color would the eyes of a fruit fly that is heterozygous for the α and β alleles be? Justify your answer.

GO ON TO THE NEXT PAGE.

7. Explain the data shown below in the graph. Include a description of the different metabolic process and their relative rates at the different parts of the graph (I, II, III, and IV).

Net Primary Productivity in a Freshwater Pond

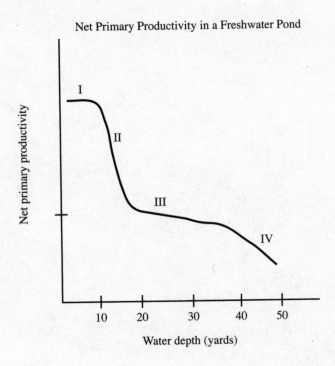

8. The human immune system involves both specific and non-specific defense mechanisms to protect against diseases caused by pathogens.

   a. Explain three types of non-specific defense mechanisms that can prevent entry and/or proliferation of a pathogen in a human.

   b. Explain three parts of the specific defense mechanisms that can help fight pathogens after initial exposure.

STOP

END OF EXAM

AP BIOLOGY PRACTICE TEST ANSWER K

Chapter 40
AP Biology
Practice Test
Answers and
Explanations

AP BIOLOGY PRACTICE TEST ANSWER KEY

1.	D
2.	D
3.	B
4.	C
5.	A
6.	C
7.	D
8.	A
9.	C
10.	B
11.	A
12.	B
13.	B
14.	D
15.	A
16.	C
17.	C
18.	C
19.	A
20.	B
21.	C
22.	B
23.	C
24.	B
25.	D
26.	C
27.	B
28.	C
29.	D
30.	C
31.	C
32.	B
33.	B
34.	C
35.	D
36.	B

37.	D
38.	B
39.	C
40.	C
41.	A
42.	A
43.	B
44.	B
45.	B
46.	C
47.	C
48.	D
49.	B
50.	D
51.	C
52.	D
53.	B
54.	B
55.	D
56.	A
57.	D
58.	C
59.	A
60.	B
61.	B
62.	B
63.	B
64	$\frac{3}{8}$
65.	0.8
66.	0.4
67.	27
68.	100
69.	25

AP BIOLOGY PRACTICE TEST ANSWERS AND EXPLANATIONS

Section I

1. **D** Absorption of emulsified lipids and fats primarily occurs in the small intestine after treatment with bile and pancreatic lipase. The primary roles of the stomach are to store food for release into the small intestine (where it is further digested and absorbed), to begin digestion of proteins (peptides) through the catalytic activity of pepsinogen (which is converted into pepsin) and hydrochloric acid (HCl) (A), the mechanical (B) and chemical breakdown of food into chyme, and disinfection of food products by treatment with HCl (C).

2. **D** Increases in substance R will result in inhibition of enzyme 2 (negative feedback). Consequently, production of P and all downstream events will be reduced such as the conversion of substance Q into substance R (D). There will be a decrease in the amount of P produced (eliminating A) because the enzyme catalyzing its production is being inhibited. Consequently, less P will be available, which will result in decreased activity of enzyme 3 (eliminating C). We would expect no appreciable change in substance M because it precedes the impacted step.

3. **B** Following depolarization of the cell to a maximum potential of +35 mV, the voltage-gated sodium (Na^+) channels will close and the voltage-gated potassium (K^+) channels will open (which causes an influx of K^+ ions and repolarization). The Na^+/K^+ pump is not voltage-dependent (rather is ATP-dependent) and remains active as long as ATP is available.

4. **C** The corpus luteum (Latin for "yellow body") is the producer of progesterone, which is responsible for readying the body for pregnancy by promoting the growth of glands and blood vessels in the endometrium.

5. **A** Divergent evolution is the process by which two species which share a common ancestry (such as dogs and humans) and evolve different structures based on differences in their behavior, traits, and respective environments. Convergent evolution, in contrast, is a process by which two unrelated species evolve similar structures based on similar selective pressures. Because these structures do not represent changes in species (speciation), the terms allopatric speciation (geographical separation causes speciation) and sympatric speciation (speciation without geographical separation).

6. **C** All messenger RNA (mRNA) strands begin with the AUG start codon. UGA is one of the three stop codons. mRNA strands, which yield cytosolic proteins, are translated by free ribosomes in the cytoplasm, whereas mRNA strands, which yield secreted proteins, are translated by ER-associated ribosomes into the lumen of the ER. Secreted proteins typically have signal peptides, which target the peptides to the ER and subsequent processing in the secretory pathway; cytosolic proteins do not need these signal sequences.

7. **D** Populations in Hardy-Weinberg equilibrium must exhibit: 1) large populations, 2) no mutations, 3) no movement into or out of the population, 4) random mating, and 5) no natural selection. Natural selection drives changes in allelic frequency and will inevitably cause changes in the gene pool (violating Hardy-Weinberg equilibrium).

8. **A** Imprinting is the learning behavior where an organism associates early on the first moving object it sees as its mother and follows and learns from it.

9. **C** Crossing-over and synapsis occurs during the first step of meiosis (prophase I). During this step, the homologous sets of chromosomes pair forming a tetrad, which permits the exchange of segments between the two sets.

10. **B** Based on the results of the data provided in the chart, at a given partial pressure of oxygen, the line associated with the most acidic pH (7.2) is also associated with the lowest oxygen saturation of hemoglobin. Therefore, as acidity increases, the oxygen saturation decreases for a given pressure of oxygen.

11. **A** Fetal hemoglobin must exhibit higher binding affinity than adult hemoglobin in order to extract oxygen from maternal blood. If there were no structural or functional difference, then the fetus would be unable to get oxygen in the womb.

12. **B** Based on the information provided, deoxygenated blood absorbs more red light and less infrared light. Therefore, the opposite must also be true; specifically, that oxygenated blood absorbs more infrared and less red light. At higher blood pH values (more basic conditions), oxygen saturation is increased for a given pressure (such as 50 mmHg in this case). Therefore, more oxygenation will occur and more infrared absorption and less red light absorption will be observed.

13. **B** Arteries, though normally carrying oxygenated blood, by definition are blood vessels that carry blood away from the heart. The pulmonary artery moves blood from the right ventricle to the lungs where it is reoxygenated. Therefore, blood in the pulmonary artery will be the least oxygenated because it reflects the point in circulation where the most oxygen has been extracted prior to cycling through the lungs.

14. **D** All blood cells, including erythrocytes, are derived from stem cells located in the bone marrow.

15. **A** Missense mutations result in a change in amino acid in the eventual protein sequence of a gene. Although missense mutations can alter or even ablate protein function, the length of the resulting protein normally remains unchanged. Nonsense mutations (B) result in a premature stop codon which will shorten the protein length. Insertions (C) and duplications (D) both result in addition of nucleotides which often cause increased protein length.

16. **C** The light-independent reactions of photosynthesis take NADPH and ATP, which have been recharged by the light-dependent reactions at the chlorophyll centers, to convert carbon dioxide into sugar.

17. **C** According to Mendelian genetics, differences in phenotypic ratios from the expected ratio are often due to linkage. Linked genes due not independently segregate resulting in less crossing over between genes, and therefore phenotypic ratios of progeny more often reflect a bias toward the parental phenotypes. Incomplete dominance (A) and codominance (B) would result in either mixing of phenotypes or full expression of multiple phenotypes, respectively. Neither of these conditions were described as an outcome. Epistatic genes (D) are dependent upon the fate of other genes. Because pea color is attributed solely to these two genes, there is not a case of epistasis.

18. **C** rRNA is the RNA which is incorporated into the structure of the ribosome and binds mRNA during translation.

19. **A** Using the data in the table, there were only two nucleotide differences between species C and D, whereas all other species had several more. Therefore, the phylogram must reflect the shortest distance between species C and D. Because A has far fewer differences from C and D, than B, it would be the next closest species.

20. **B** Inhibitor Y is binding in a site that is distant from the active site which still causes impairment of function. This type of inhibition is called allosteric regulation. Because the substrate must impair enzymatic activity, it likely induces a conformational change which either limits access to the active site or prevents catalytic activity after substrate binding.

21. **C** Photosynthesis occurs in chloroplasts (not mitochondria) so you can eliminate (A) right off the bat. Light penetrates the chloroplast and strikes antenna complexes in thylakoids to activate chlorophyll during the light-dependent cycles. The numerous stacks of thylakoids in chloroplasts are called grana.

22. **B** Methionine and cysteine are amino acids which are normally incorporated into growing peptide chains during protein translation. The assembly of bacteriophages will result in the translation and incorporation of protein into the protein capsid coat of the virus. The genome and envelope (if present) of the phage are likely to have very little protein and therefore very little amino acid incorporation. Viruses don't have nuclei, so there would be no labeling of the nuclear envelope.

23. **C** During transcription, adenine (A) bases are paired with uracil (U) bases, and cytosine (C) bases are paired with guanine (G) bases. There is no thymine in RNA, eliminating (A) and (B). The only sequence that accurately pairs bases in the correct orientation is answer (C).

24. **B** Wind-blown forests of conifers describes the conditions of the biome taiga. If the fields become more bare and acquire a permafrost, they are transitioning to a more arctic climate consistent with the biome tundra.

25. **D** Sacrificial behavior for the benefit of the colony or population is referred to as altruistic behavior.

26. **C** Based on the data shown, strain C is able to grow in the presence of all four antibiotics, whereas both strains A and B are sensitive to at least one type of antibiotic. Therefore, strain C would be most resistant.

27. **B** Kanamycin prevented growth of colonies of strain B indicative of bacterial sensitivity. The patient should be treated with kanamycin.

28. **C** Based on the results of the follow-up experiment, the bacterial strains only vary in expression of antibiotic resistance genes. The presence of a unique flagellum and attachment proteins would indicate the presence of additional genes that are not associated with antibiotic resistance. This finding would not support the hypothesis.

29. **D** Bacteria lack membrane-bound organelles, which includes nuclei. The primary structures of bacteria are a cell wall, plasma membrane, ribosomes, nucleoid region (where the DNA aggregates), and possible motility and attachment structures (such as fimbriae, a capsule, or flagella).

30. **C** Transformation is the process by which bacteria acquire extracellular free-floating DNA from their environment. Because no phages are present, transduction cannot occur. Transfection and PCR describe processes that are not related to bacterial DNA acquisition.

31. **C** Catalysts accelerate reactions by lowering the activation energy (the energy threshold that must be met for a reaction to proceed to products). Catalysts have no effect on the energy of the reactants or products.

32. **B** Trypsin and chymotrypsin are digestive enzymes which breakdown proteins. The only enzyme listed that also breaks down proteins is pepsin. Lipase breaks down lipids (or fats). Amylase and lactase break down sugars.

33. **B** B cells are responsible for producing antibodies and mediating humoral immunity to antigens (or pathogens). Phagocytes are innate immune cells which engulf non-self pathogens by phagocytosis. Cytotoxic T cells destroy infected cells. Helper T cells trigger the immune response to sites of infection.

34. **C** The liver is not a part of the gastrointestinal tract. Instead, it releases bile salts via the common bile duct into the duodenum of the small intestine and as such is considered an accessory organ.

35. **D** Selection towards one extreme or the other is referred to as directional selection.

36. **B** Because prairie dogs are directly consuming the producers, they are considered the primary consumers in this community.

37. **D** The ectoderm germ layer becomes the skin and nervous system following organogenesis.

38. **B** Smooth muscle and cardiac cells have many similarities, which include a single nucleus (in contrast the multinucleated cells of skeletal muscles) and they are both regulated by involuntary control. However, only cardiac and skeletal muscle cells are striated, not smooth muscles (hence the name).

39. **C** Growth hormone is produced by the pituitary gland. Gigantism is commonly caused by excessive production (usually by a tumor) of growth hormone.

40. **C** Chromosomes are separated during anaphase. The cell hasn't yet entered telophase because no nuclear envelopes are detected and the cleavage furrow hasn't formed.

41. **A** The inhibitor must be blocking an active process of anaphase. During anaphase, the microtubules contract pulling the chromatids to the poles of the cells. Inhibition of this process would arrest the chromatids in the process of separation. Mitotic spindles have already formed as evidenced by the connected microtubules.

42. **A** Because the cell is undergoing mitosis, it isn't likely a spermatocyte (which undergoes meiosis during spermatogenesis). Hepatocytes (B), lymphocytes (C), and epithelial cells (D) are all somatic cells and undergo mitosis.

43. **B** Upon degradation of the inhibitor, the cell is expected to enter telophase. During telophase, the nuclear envelope will reappear in each of the two daughter cell segments as cytokinesis occurs.

44. **B** Alpha helices are part of the secondary structure of proteins. Secondary structure is due to backbone folding interactions and limitations.

45. **B** Nondisjunction is the abnormal separation of chromosomes (or chromatids during meiosis II) during meiosis. If the chromosomes do not segregate evenly between the cells, then one cell can receive two copies of a chromosome. During fertilization, the addition of a third copy results in trisomy (or three chromosome copies). For most chromosomes, trisomy is lethal. However, in the case of Trisomy 21, the condition isn't lethal but is associated with clear physical and mental impairments.

46. **C** The father must be heterozygous (carrying one A and one O allele) because the older sibling of the child in question is O. Had the father been homozygous, then the sibling could not be O. Therefore, there is a 1 out of 2 chance that the father passes on the A allele, which being dominant, would result in an A blood type.

47. **C** The alveoli of the lungs are the terminal structures of the respiratory tract where gas is exchanged with adjacent capillaries.

48. **D** Aldosterone causes increased absorption of sodium ions from the filtrate in the distal convoluted tubule (DCT).

49. **B** Antidiuretic hormone (ADH) is released to increase water reabsorption in the collecting duct in response to low blood volume or pressure. Inhibitors of ADH, such as coffee, block this activity, resulting in decreased water reabsorption and therefore increased urine production.

50. **D** Albumin is a very common blood associated protein. Proteins are too large to be filtered out into Bowman's capsule under normal conditions. The presence of protein in the urine (proteinuria) is a sign of kidney failure or disease. Salt (A), water (B), and urea (C) are all filtered out of the blood in the nephron.

51. **C** Urea is a byproduct of protein catabolism. As proteins are broken down, they are either converted into ammonia or are directly converted to urea. Urea is more stable and less toxic than ammonia and is generated to safely remove excess nitrogen wastes from the body.

52. **D** At the peak potential (around +35 mV), voltage-gated sodium channels close and voltage-gated potassium channels open. Because there is more potassium inside the cell than outside, the potassium ions rush out of the cell causing a repolarization (drop in potential).

53. **B** Helicase unwinds the two strands of the double helix in preparation for DNA replication.

54. **B** Chlorofluorocarbons (CFCs) have been known for many years now to cause degradation of ozone molecules in the atmosphere. Their phase out is meant to reduce the effects of ozone depletion.

55. **D** Thyroid-stimulating hormone (TSH) is released by the pituitary gland to trigger production and secretion of the thyroid hormone, thyroxine.

56. **A** Hearing and sight are coordinated by the auditory and occipital regions of the cerebrum. Damage to these regions has most resulted in perceived hearing loss and blindness.

57. **D** There is a $\frac{1}{2}$ chance of a gamete receiving each of the individual alleles. Therefore, the odds that a gamete would receive all four of the recessive alleles is equal to $\left(\frac{1}{2}\right)^4$ or $\frac{1}{16}$.

58. **C** Using the scale of ecological succession shown, the forest is likely between the white pine-spruce sere and the start of the beech-maple sere.

59. **A** Following forest fires, the secondary community normally grows much faster than the initial community.

60. **B** Although both amylases perform similar functions, the salivary amylase is denatured in the presence of low pH and digested by pepsin. However, digestion of carbohydrates is normally not complete by the time that the bolus reaches the stomach. Therefore, pancreatic amylase is secreted to continue digestion after chyme is released into the duodenum.

61. **B** Hypertonic solutions have more salt than the intracellular environment of the cell in which they bathe. Therefore, water undergoes osmosis from inside the cell to the outside environment causing it to shrink. An example of this process is pickles, which are bathed in a brine (salt solution).

62. **B** Parathyroid hormone (PTH) is released in response to low serum concentrations of calcium to increase calcium levels by catalyzing the breakdown of bone and increasing calcium absorption in the kidneys and intestines.

63. **B** Humans produce lactic acid by fermentation, whereas yeast produce ethanol. The ethanol fermentation process used by yeast is what generates beer and wine.

64. $\dfrac{3}{8}$ The cross in this process is FfPp × ffPp. There is a 50/50 chance that the progeny will have purple flowers (receiving the F allele), and there is a $\dfrac{3}{4}$ chance that the progeny will have green pods (receiving the P allele). Therefore, the total fraction having both dominant alleles is $\dfrac{1}{2} \times \dfrac{3}{4}$ or $\dfrac{3}{8}$.

65. **0.8** Because 4% of the population has dry earwax, they must be homozygous for the allele (let us use q for the recessive allele) or $q^2 = 0.04$. Therefore, the frequency of the recessive allele must be the square root of 0.04 or 0.2. Using the formula, $p + q = 1$, the frequency of the dominant allele must be 0.8.

66. **0.4** Using the chart provided, the population in 1450 appears to be halfway between 800,000 and 0. Therefore, it is most nearly 400,000 or 0.4 million people.

67. **27** Energy must be conserved; therefore if plankton receive 60 g/m² and lose 33 g/m² to primary consumers and detritus, then they must respire the remaining 27 g/m².

68. **100** The stationary point of the curve shows a titer for wild-type virus of 10^6 (6 logs) and 10^4 (4 logs) for the *ts* mutant. Therefore, the wild-type virus exhibits a 100-fold greater titer (2 logs) over the mutant.

69. **25** There is a 50% chance that the couple will have a boy. Because the boy would have to receive his Y chromosome from his father, the only chance that he will have hemophilia will come from his mother. The child's mother must be a carrier for the disease because she had to receive her father's diseased X chromosome. Therefore, there is a 50/50 chance that the boy will receive the hemophilia allele from his mother. Multiplying the odds of receiving the disease allele (1/2) by the odds of having a boy (1/2) results in a 25% chance of the couple having an afflicted boy.

Section II

1. a. This is an example of pedigree chart for a family of wolves impacted by a X-linked disease. Because only male wolves are affected, the disease is associated with the X chromosome. We know that this is not a Y-linked disease because not all males are affected.

   b. There is a 25% chance that the offspring of the cross between xvii and xii will result in a diseased female pup. First, the father is carrying a diseased allele (X'Y), and the mother has a 50/50 chance of having a diseased allele (X'X) because her mother is a carrier (we know this because her sibling is diseased). The only way that a female pup can have the disease is if both alleles are diseased. There is a 100% chance that she will receive a diseased allele from her father and a 50% chance from her mother (if she is a carrier). Therefore, the odds are $\frac{1}{2}$ (mother is a carrier) $\times$ $\frac{1}{2}$ (she gets a bad copy from her mother) or $\frac{1}{4}$.

c.

	X*	Y
X	X*X	XY
X	X*X	XY

The diseased father of the pups cannot pass on his diseased allele to his children because they are both male and their mother is assumed to be disease-free because she bred into the family. Therefore, the males should not have the disease. All female pups born will be carriers for the disease because they must receive their father's diseased X chromosome.

d. If the frequency of the diseased allele is 0.1, then the frequency of affected individuals would be $(0.1)^2$ or 0.01 (1%).

e. If neither of the parents for both wolves showed no disease, then the most likely cause of the disease was a random mutation in the mother.

2. a.

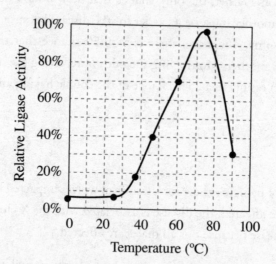

At temperatures above 75° C, the enzyme activity degrades sharply. The enzyme is likely being denatured at the higher temperatures.

b. At concentrations above 0.5 μM, the enzyme is likely saturated with substrate and is working at its maximum activity. Addition of more enzymes has no effect, because the substrate is the limiting reagent.

c.

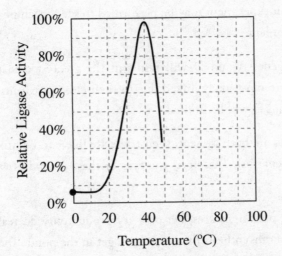

Because the enzyme activity appears to correlate with the physiological conditions of the host, we would expect the enzyme to function to an optimum of approximately 37° C. Increases in temperature above this would likely result in abrupt drop in activity assuming the hypothesis is correct.

d.   The ligases of *Pyrospirallis* and *Thermoacidophilus* share similar ligase optimum activities to that of the discovered ligase. Because both of these organisms thrive in deep sea thermal vents. It is very likely that this organism may be able to thrive in a similar environment.

3.   a.   In the last few hundred years, the kakapo have experienced a high selective pressure against birds that are unable to leap or escape from cats. Therefore, a likely hypothesis is that the increase in cat populations has caused provided a selective advantage for birds with increased leaping and gliding abilities.

b.   There are many experiments which could be performed to evaluate this phenomenon. PCR or genetic analysis can be performed on old kakapo samples and new kakapo samples to evaluate the relative amounts of genes associated with leaping or gliding. An additional test would be to place kakapo birds in environments free of cats and see if there is a reduction (or return) to historic leaping or gliding abilities.

4.   In the curve, the bacterial culture undergoes exponential growth between the hours of 2 and 5 due to high availability of resources and uncontrolled expansion. At 5 hours, the culture reaches a stationary phase where the 1 L flask has reached a carrying capacity (limitation of resources and space for continued expansion). By 9 hours, the amount of resources has declined to the point that the carrying capacity can no longer be maintained. At these conditions, there is more bacterial death than growth.

5. There are many answers which may fit this question. First, genetic data (sequences) can be collected to compare the different populations of algae to see if they are the same species or are closely related. Secondly, an experiment may be performed to see if each population is capable of growing in the other environment.

6. a. Crosses involving genes A and D will result in the greatest deviation from the expected ratio because the genes are most likely linked due to their close proximity. Consequently, crossing over events between A and D will be limited.

   b. If the α and β alleles are both present, their traits are likely to be mixed (as is the case with incomplete dominance scenarios). Therefore, the eyes of these fruit flies would likely be some shade of purple.

7. As the depth in the pond increases, net primary productivity decreases, as there is less light available and thus, less photosynthesis the deeper you get in the pond. The area of the graph labeled I is the most shallow part of the pond, and it is the most productive because light is not a limiting factor at the surface of the pond, so there is no limitation to photosynthesis. At area II, there is a rapid decrease in the amount of productivity as there is decreasing light available for photosynthesis as you get deeper in the pond. Around 0 on the y-axis, in area III (the compensation point, where production and consumption are equal), the photosynthetic product is equal to the cellular respiration requirements for the organisms at the level in the pond. Finally, area IV shows that below 0, cellular respiration requirements outpace the photosynthetic product, so there is an increasingly negative net primary productivity as you get deeper in the pond.

8. a. Three of any of the following: skin as a barrier, mucus/hair/ear wax to trap pathogens, white blood cells that can undergo phagocytosis, elimination of pathogens via coughing/sneezing/urination, high pH in stomach acid/urine to kill pathogens, destruction of pathogens by complement or natural killer cells, lysozyme (enzyme that kills bacteria) in tears/sweat, general inflammatory response (increase in temperature, histamine release, vasocilation, macrophage recruitment).

   b. Three of any of the following: antigen presenting cells (macrophages, dendritic cells, B cells) presenting antigen to B cells and T cells, B cells producing antibodies, helper T cells activate B cells, killer T cells kill abnormal cells via apoptosis, cytokines for signaling/activation, memory cells produced during primary exposure to speed up secondary response.

About the Authors

Christopher C. Stobart, Ph.D. received his Bachelor of Science degrees in Biology and Chemistry from Xavier University and his Ph.D. in Microbiology and Immunology from Vanderbilt University. Since 2008, Chris has worked with countless students in The Princeton Review prep courses and through one-on-one tutoring for a variety of tests including several AP exams, the ACT, MCAT, and DAT. Chris is currently working as a postdoctoral research fellow at Emory University in Atlanta and enjoys spending his free time traveling and exploring the outdoors with his wife.

Mary DeAgostino-Kelly is a medical student at Vanderbilt University and plans to go into OB/GYN. She currently teaches SAT and MCAT classes for The Princeton Review, and enjoys spending time with her husband, Kevin, and dog, Rocky, in their adopted home of Nashville, TN.

Sarah Abigail Woodruff received her Bachelor of Science degree in microbiology and Bachelor of Arts degree in Women's Studies from the University of Maryland and her Bachelor of Science degree in mathematics and statistics from Georgia State University. She has worked with The Princeton Review since 2001 and specializes in MCAT biology, SAT biology, AP biology and GRE. Her roles at home include spouse, mother to a daughter and caretaker for a small menagerie of animals.

The Princeton Review

1. YOUR NAME:
(Print)
Last   First   M.I.

SIGNATURE: DATE: /   /

HOME ADDRESS:
(Print)
Number and Street

City   State   Zip Code

PHONE NO. :
(Print)

5. YOUR NAME

First 4 letters of last name				FIRST INIT	MID INIT

A B C D E F G H I J K L M N O P Q R S T U V W X Y Z

IMPORTANT: Please fill in these boxes exactly as shown on the back cover of your test book.

2. TEST FORM

3. TEST CODE   **4. REGISTRATION NUMBER**

0 1 2 3 4 5 6 7 8 9

A B C D E F G

6. DATE OF BIRTH

Month	Day	Year
JAN		
FEB		
MAR	0 0	0 0
APR	1 1	1 1
MAY	2 2	2 2
JUN	3 3	3 3
JUL	4	4 4
AUG	5	5 5
SEP	6	6 6
OCT	7	7 7
NOV	8	8 8
DEC	9	9 9

7. SEX
MALE
FEMALE

The Princeton Review
© The Princeton Review, Inc.
FORM NO. 00001-PR

Section 1   Start with number 1 for each new section.
If u section has fewer questions than answer spaces, leave the extra answer spaces blank.

1. A B C D   31. A B C D   61. A B C D   91. A B C D
2. A B C D   32. A B C D   62. A B C D   92. A B C D
3. A B C D   33. A B C D   63. A B C D   93. A B C D
4. A B C D   34. A B C D   64. A B C D   94. A B C D
5. A B C D   35. A B C D   65. A B C D   95. A B C D
6. A B C D   36. A B C D   66. A B C D   96. A B C D
7. A B C D   37. A B C D   67. A B C D   97. A B C D
8. A B C D   38. A B C D   68. A B C D   98. A B C D
9. A B C D   39. A B C D   69. A B C D   99. A B C D
10. A B C D   40. A B C D   70. A B C D   100. A B C D
11. A B C D   41. A B C D   71. A B C D   101. A B C D
12. A B C D   42. A B C D   72. A B C D   102. A B C D
13. A B C D   43. A B C D   73. A B C D   103. A B C D
14. A B C D   44. A B C D   74. A B C D   104. A B C D
15. A B C D   45. A B C D   75. A B C D   105. A B C D
16. A B C D   46. A B C D   76. A B C D   106. A B C D
17. A B C D   47. A B C D   77. A B C D   107. A B C D
18. A B C D   48. A B C D   78. A B C D   108. A B C D
19. A B C D   49. A B C D   79. A B C D   109. A B C D
20. A B C D   50. A B C D   80. A B C D   110. A B C D
21. A B C D   51. A B C D   81. A B C D   111. A B C D
22. A B C D   52. A B C D   82. A B C D   112. A B C D
23. A B C D   53. A B C D   83. A B C D   113. A B C D
24. A B C D   54. A B C D   84. A B C D   114. A B C D
25. A B C D   55. A B C D   85. A B C D   115. A B C D
26. A B C D   56. A B C D   86. A B C D   116. A B C D
27. A B C D   57. A B C D   87. A B C D   117. A B C D
28. A B C D   58. A B C D   88. A B C D   118. A B C D

. YOUR NAME: _____
(Print)   Last   First   M.I.

IGNATURE: _____ **DATE:** _____ / _____ / _____

OME ADDRESS: _____
(Print)   Number and Street

                City   State   Zip Code

PHONE NO. : _____
(Print)

MPORTANT: Please fill in these boxes exactly as shown on the back cover of your test book.

5. YOUR NAME

First 4 letters of last name				FIRST INIT	MID INIT
Ⓐ	Ⓐ	Ⓐ	Ⓐ	Ⓐ	Ⓐ
Ⓑ	Ⓑ	Ⓑ	Ⓑ	Ⓑ	Ⓑ
Ⓒ	Ⓒ	Ⓒ	Ⓒ	Ⓒ	Ⓒ
Ⓓ	Ⓓ	Ⓓ	Ⓓ	Ⓓ	Ⓓ
Ⓔ	Ⓔ	Ⓔ	Ⓔ	Ⓔ	Ⓔ
Ⓕ	Ⓕ	Ⓕ	Ⓕ	Ⓕ	Ⓕ
Ⓖ	Ⓖ	Ⓖ	Ⓖ	Ⓖ	Ⓖ
Ⓗ	Ⓗ	Ⓗ	Ⓗ	Ⓗ	Ⓗ
Ⓘ	Ⓘ	Ⓘ	Ⓘ	Ⓘ	Ⓘ
Ⓙ	Ⓙ	Ⓙ	Ⓙ	Ⓙ	Ⓙ
Ⓚ	Ⓚ	Ⓚ	Ⓚ	Ⓚ	Ⓚ
Ⓛ	Ⓛ	Ⓛ	Ⓛ	Ⓛ	Ⓛ
Ⓜ	Ⓜ	Ⓜ	Ⓜ	Ⓜ	Ⓜ
Ⓝ	Ⓝ	Ⓝ	Ⓝ	Ⓝ	Ⓝ
Ⓞ	Ⓞ	Ⓞ	Ⓞ	Ⓞ	Ⓞ
Ⓟ	Ⓟ	Ⓟ	Ⓟ	Ⓟ	Ⓟ
Ⓠ	Ⓠ	Ⓠ	Ⓠ	Ⓠ	Ⓠ
Ⓡ	Ⓡ	Ⓡ	Ⓡ	Ⓡ	Ⓡ
Ⓢ	Ⓢ	Ⓢ	Ⓢ	Ⓢ	Ⓢ
Ⓣ	Ⓣ	Ⓣ	Ⓣ	Ⓣ	Ⓣ
Ⓤ	Ⓤ	Ⓤ	Ⓤ	Ⓤ	Ⓤ
Ⓥ	Ⓥ	Ⓥ	Ⓥ	Ⓥ	Ⓥ
Ⓦ	Ⓦ	Ⓦ	Ⓦ	Ⓦ	Ⓦ
Ⓧ	Ⓧ	Ⓧ	Ⓧ	Ⓧ	Ⓧ
Ⓨ	Ⓨ	Ⓨ	Ⓨ	Ⓨ	Ⓨ
Ⓩ	Ⓩ	Ⓩ	Ⓩ	Ⓩ	Ⓩ

2. TEST FORM

6. DATE OF BIRTH

Month	Day	Year
◯ JAN		
◯ FEB		
◯ MAR	Ⓞ Ⓞ	Ⓞ Ⓞ
◯ APR	① ①	① ①
◯ MAY	② ②	② ②
◯ JUN	③ ③	③ ③
◯ JUL	④	④ ④
◯ AUG	⑤	⑤ ⑤
◯ SEP	⑥	⑥ ⑥
◯ OCT	⑦	⑦ ⑦
◯ NOV	⑧	⑧ ⑧
◯ DEC	⑨	⑨ ⑨

3. TEST CODE   4. REGISTRATION NUMBER

Ⓞ Ⓐ Ⓞ Ⓞ Ⓞ | Ⓞ Ⓞ Ⓞ Ⓞ Ⓞ Ⓞ
① Ⓑ ① ① ① | ① ① ① ① ① ①
② Ⓒ ② ② ② | ② ② ② ② ② ②
③ Ⓓ ③ ③ ③ | ③ ③ ③ ③ ③ ③
④ Ⓔ ④ ④ ④ | ④ ④ ④ ④ ④ ④
⑤ Ⓕ ⑤ ⑤ ⑤ | ⑤ ⑤ ⑤ ⑤ ⑤ ⑤
⑥ Ⓖ ⑥ ⑥ ⑥ | ⑥ ⑥ ⑥ ⑥ ⑥ ⑥
⑦ ⑦ ⑦ ⑦ | ⑦ ⑦ ⑦ ⑦ ⑦ ⑦
⑧ ⑧ ⑧ ⑧ | ⑧ ⑧ ⑧ ⑧ ⑧ ⑧
⑨ ⑨ ⑨ ⑨ | ⑨ ⑨ ⑨ ⑨ ⑨ ⑨

7. SEX
◯ MALE
◯ FEMALE

The Princeton Review
© The Princeton Review, Inc.
FORM NO. 00001-PR

Section ①

Start with number 1 for each new section.
If a section has fewer questions than answer spaces, leave the extra answer spaces blank.

1. Ⓐ Ⓑ Ⓒ Ⓓ
2. Ⓐ Ⓑ Ⓒ Ⓓ
3. Ⓐ Ⓑ Ⓒ Ⓓ
4. Ⓐ Ⓑ Ⓒ Ⓓ
5. Ⓐ Ⓑ Ⓒ Ⓓ
6. Ⓐ Ⓑ Ⓒ Ⓓ
7. Ⓐ Ⓑ Ⓒ Ⓓ
8. Ⓐ Ⓑ Ⓒ Ⓓ
9. Ⓐ Ⓑ Ⓒ Ⓓ
10. Ⓐ Ⓑ Ⓒ Ⓓ
11. Ⓐ Ⓑ Ⓒ Ⓓ
12. Ⓐ Ⓑ Ⓒ Ⓓ
13. Ⓐ Ⓑ Ⓒ Ⓓ
14. Ⓐ Ⓑ Ⓒ Ⓓ
15. Ⓐ Ⓑ Ⓒ Ⓓ
16. Ⓐ Ⓑ Ⓒ Ⓓ
17. Ⓐ Ⓑ Ⓒ Ⓓ
18. Ⓐ Ⓑ Ⓒ Ⓓ
19. Ⓐ Ⓑ Ⓒ Ⓓ
20. Ⓐ Ⓑ Ⓒ Ⓓ
21. Ⓐ Ⓑ Ⓒ Ⓓ
22. Ⓐ Ⓑ Ⓒ Ⓓ
23. Ⓐ Ⓑ Ⓒ Ⓓ
24. Ⓐ Ⓑ Ⓒ Ⓓ
25. Ⓐ Ⓑ Ⓒ Ⓓ
26. Ⓐ Ⓑ Ⓒ Ⓓ
27. Ⓐ Ⓑ Ⓒ Ⓓ
28. Ⓐ Ⓑ Ⓒ Ⓓ

31. Ⓐ Ⓑ Ⓒ Ⓓ
32. Ⓐ Ⓑ Ⓒ Ⓓ
33. Ⓐ Ⓑ Ⓒ Ⓓ
34. Ⓐ Ⓑ Ⓒ Ⓓ
35. Ⓐ Ⓑ Ⓒ Ⓓ
36. Ⓐ Ⓑ Ⓒ Ⓓ
37. Ⓐ Ⓑ Ⓒ Ⓓ
38. Ⓐ Ⓑ Ⓒ Ⓓ
39. Ⓐ Ⓑ Ⓒ Ⓓ
40. Ⓐ Ⓑ Ⓒ Ⓓ
41. Ⓐ Ⓑ Ⓒ Ⓓ
42. Ⓐ Ⓑ Ⓒ Ⓓ
43. Ⓐ Ⓑ Ⓒ Ⓓ
44. Ⓐ Ⓑ Ⓒ Ⓓ
45. Ⓐ Ⓑ Ⓒ Ⓓ
46. Ⓐ Ⓑ Ⓒ Ⓓ
47. Ⓐ Ⓑ Ⓒ Ⓓ
48. Ⓐ Ⓑ Ⓒ Ⓓ
49. Ⓐ Ⓑ Ⓒ Ⓓ
50. Ⓐ Ⓑ Ⓒ Ⓓ
51. Ⓐ Ⓑ Ⓒ Ⓓ
52. Ⓐ Ⓑ Ⓒ Ⓓ
53. Ⓐ Ⓑ Ⓒ Ⓓ
54. Ⓐ Ⓑ Ⓒ Ⓓ
55. Ⓐ Ⓑ Ⓒ Ⓓ
56. Ⓐ Ⓑ Ⓒ Ⓓ
57. Ⓐ Ⓑ Ⓒ Ⓓ
58. Ⓐ Ⓑ Ⓒ Ⓓ

61. Ⓐ Ⓑ Ⓒ Ⓓ
62. Ⓐ Ⓑ Ⓒ Ⓓ
63. Ⓐ Ⓑ Ⓒ Ⓓ
64. Ⓐ Ⓑ Ⓒ Ⓓ
65. Ⓐ Ⓑ Ⓒ Ⓓ
66. Ⓐ Ⓑ Ⓒ Ⓓ
67. Ⓐ Ⓑ Ⓒ Ⓓ
68. Ⓐ Ⓑ Ⓒ Ⓓ
69. Ⓐ Ⓑ Ⓒ Ⓓ
70. Ⓐ Ⓑ Ⓒ Ⓓ
71. Ⓐ Ⓑ Ⓒ Ⓓ
72. Ⓐ Ⓑ Ⓒ Ⓓ
73. Ⓐ Ⓑ Ⓒ Ⓓ
74. Ⓐ Ⓑ Ⓒ Ⓓ
75. Ⓐ Ⓑ Ⓒ Ⓓ
76. Ⓐ Ⓑ Ⓒ Ⓓ
77. Ⓐ Ⓑ Ⓒ Ⓓ
78. Ⓐ Ⓑ Ⓒ Ⓓ
79. Ⓐ Ⓑ Ⓒ Ⓓ
80. Ⓐ Ⓑ Ⓒ Ⓓ
81. Ⓐ Ⓑ Ⓒ Ⓓ
82. Ⓐ Ⓑ Ⓒ Ⓓ
83. Ⓐ Ⓑ Ⓒ Ⓓ
84. Ⓐ Ⓑ Ⓒ Ⓓ
85. Ⓐ Ⓑ Ⓒ Ⓓ
86. Ⓐ Ⓑ Ⓒ Ⓓ
87. Ⓐ Ⓑ Ⓒ Ⓓ
88. Ⓐ Ⓑ Ⓒ Ⓓ

91. Ⓐ Ⓑ Ⓒ Ⓓ
92. Ⓐ Ⓑ Ⓒ Ⓓ
93. Ⓐ Ⓑ Ⓒ Ⓓ
94. Ⓐ Ⓑ Ⓒ Ⓓ
95. Ⓐ Ⓑ Ⓒ Ⓓ
96. Ⓐ Ⓑ Ⓒ Ⓓ
97. Ⓐ Ⓑ Ⓒ Ⓓ
98. Ⓐ Ⓑ Ⓒ Ⓓ
99. Ⓐ Ⓑ Ⓒ Ⓓ
100. Ⓐ Ⓑ Ⓒ Ⓓ
101. Ⓐ Ⓑ Ⓒ Ⓓ
102. Ⓐ Ⓑ Ⓒ Ⓓ
103. Ⓐ Ⓑ Ⓒ Ⓓ
104. Ⓐ Ⓑ Ⓒ Ⓓ
105. Ⓐ Ⓑ Ⓒ Ⓓ
106. Ⓐ Ⓑ Ⓒ Ⓓ
107. Ⓐ Ⓑ Ⓒ Ⓓ
108. Ⓐ Ⓑ Ⓒ Ⓓ
109. Ⓐ Ⓑ Ⓒ Ⓓ
110. Ⓐ Ⓑ Ⓒ Ⓓ
111. Ⓐ Ⓑ Ⓒ Ⓓ
112. Ⓐ Ⓑ Ⓒ Ⓓ
113. Ⓐ Ⓑ Ⓒ Ⓓ
114. Ⓐ Ⓑ Ⓒ Ⓓ
115. Ⓐ Ⓑ Ⓒ Ⓓ
116. Ⓐ Ⓑ Ⓒ Ⓓ
117. Ⓐ Ⓑ Ⓒ Ⓓ
118. Ⓐ Ⓑ Ⓒ Ⓓ

NOTES

NOTES

NOTES

NOTES

NOTES

NOTES

NOTES

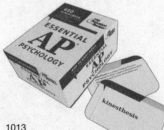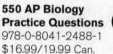